中国热区植原体病害多样性及其诊断与防控

◎ 于少帅　车海彦　潘英文　编著

ZHONGGUO REQU
ZHIYUANTI BINGHAI DUOYANGXING
JIQI ZHENDUAN YU FANGKONG

中国农业科学技术出版社

图书在版编目（CIP）数据

中国热区植原体病害多样性及其诊断与防控 / 于少帅，车海彦，潘英文编著 . -- 北京：中国农业科学技术出版社，2024. 11. -- ISBN 978-7-5116-7161-5

Ⅰ . S432

中国国家版本馆 CIP 数据核字第 202465Z2Z7 号

责任编辑　姚　欢
责任校对　王　彦
责任印制　姜义伟　王思文

出 版 者	中国农业科学技术出版社
	北京市中关村南大街 12 号　　邮编：100081
电　　话	（010）82106631（编辑室）　（010）82106624（发行部）
	（010）82109709（读者服务部）
网　　址	https：// castp.caas.cn
经 销 者	各地新华书店
印 刷 者	中煤（北京）印务有限公司
开　　本	185 mm × 260 mm　1/16
印　　张	12.5
字　　数	300 千字
版　　次	2024 年 11 月第 1 版　2024 年 11 月第 1 次印刷
定　　价	128.00 元

版权所有·侵权必究

《中国热区植原体病害多样性及其诊断与防控》编委会

主 编 著：于少帅，中国热带农业科学院椰子研究所
　　　　　车海彦，中国热带农业科学院环境与植物保护研究所
　　　　　潘英文，海口海关热带植物隔离检疫中心

编者成员（按姓氏笔画排序）：
　　　　　于少帅，中国热带农业科学院椰子研究所
　　　　　车海彦，中国热带农业科学院环境与植物保护研究所
　　　　　王圣洁，中国林业科学研究院热带林业研究所
　　　　　李正男，内蒙古农业大学
　　　　　朱　辉，中国热带农业科学院椰子研究所
　　　　　宋传生，菏泽学院
　　　　　宋薇薇，中国热带农业科学院椰子研究所
　　　　　杨德洁，中国热带农业科学院椰子研究所
　　　　　张　凤，湖南农业大学
　　　　　张　磊，内蒙古农业大学
　　　　　吴　华，海南医科大学
　　　　　林兆威，中国热带农业科学院椰子研究所
　　　　　潘英文，海口海关热带植物隔离检疫中心

前言 PREFACE

植原体是一类寄生于植物和昆虫、无细胞壁、尚难分离培养的原核致病菌，可侵染1 000多种植物，在全球范围内造成严重的经济损失和生态破坏。我国热区植原体多样性丰富，已报道植原体病害种类120余种，占我国已报道植原体病害的60%以上。植原体病害给我国热区农业生产与生态环境带来严重影响，如槟榔黄化病、花生丛枝病、水稻橙叶病等。本书基于作者在我国热区植物植原体病害领域的研究及取得的成果，系统介绍了我国热区植原体病害多样性及其诊断与防控技术，阐明了我国热区植原体遗传多样性及其病害检测鉴定、监测预警、发生流行、防控管理等方面的特征及研究进展，为本领域科学研究、技术研发、推广应用等提供理论参考与技术支撑。

本书共8章。第一章和第二章概述了植原体生物学特征、系统分类研究现状等；第三章系统介绍了植原体及其病害检测诊断技术；第四章介绍了植原体与寄主植物互作的研究现状及植原体的致病机理；第五章系统描述了我国热区植原体遗传多样性及其植原体病害田间典型症状、地理分布、传播方式等；第六章系统介绍了植原体与病毒、韧皮部杆菌、螺原体等病原微生物复合侵染引起的植物病害现象；第七章和第八章介绍了植原体病害的田间防控管理，并对该类病害研究的发展趋势进行了展望。

本书的出版得到了中央级公益性科研院所基本科研业务费专项（No. 1630152024009）的资助，相关研究工作得到了海南省自然科学基金高层次人才项目（No. 320RC743、No. 323RC524）、中央级公益性科研院所基本科研业务费专项（No. 1630152021005、No. 1630152022004、No. 1630042017023）、海南省科学技术厅农业科技成果转化资金项目（No. 2014GB2E200114）、海南省重大科技计划项目（No. ZDKJ201817）、海南省院士创新平台专项（No. YSPTZX202138）等项目的支持，在此一并致谢！

由于作者水平有限，书中难免存在不足之处，敬请相关领域的专家学者及广大读者在阅读和使用过程中提出宝贵意见和建议，以便重印或再版时修订和完善。

于少帅　车海彦　潘英文
2024年6月

目 录 / CONTENTS

第一章　植原体研究概述 … 1
一、植原体生物学特征 … 1
二、植原体基因组 … 2
三、我国热区植原体遗传多样性 … 6
参考文献 … 7

第二章　植原体系统分类 … 13
一、植原体16Sr组 … 13
二、植原体候选种 … 14
参考文献 … 24

第三章　植原体及其病害检测诊断技术 … 33
一、电子显微镜检测技术 … 33
二、组织化学技术 … 33
三、PCR扩增技术 … 34
四、巢式PCR扩增技术 … 35
五、实时荧光定量PCR技术 … 36
六、环介导等温基因扩增技术 … 37
七、微滴式数字PCR扩增技术 … 48
八、微阵列生物芯片检测技术 … 52
九、多位点检测技术 … 56
十、核酸杂交检测技术 … 65
十一、免疫检测技术 … 66
参考文献 … 66

第四章　植原体致病机理 …… 73
一、植原体效应子 …… 73
二、植原体效应子功能及其致病机理 …… 75
三、植原体启动子结构与功能 …… 78
参考文献 …… 87

第五章　植原体病害多样性 …… 92
一、林木植物 …… 92
二、粮食作物 …… 105
三、油料作物 …… 109
四、糖料作物 …… 110
五、纤维作物 …… 111
六、果树作物 …… 113
七、蔬菜作物 …… 117
八、香辛料作物 …… 126
九、饲料及绿肥作物 …… 128
十、观赏植物 …… 131
十一、其他植物 …… 137
参考文献 …… 151

第六章　植原体的复合侵染 …… 164
一、不同组植原体的复合侵染 …… 164
二、植原体与病毒复合侵染 …… 164
三、植原体与柑橘黄化病菌复合侵染 …… 165
四、植原体与马铃薯斑纹片病菌复合侵染 …… 166
五、植原体与螺原体复合侵染 …… 167
六、植原体与其他难培养病原菌复合侵染 …… 167
七、植原体与真菌复合侵染 …… 167
参考文献 …… 168

第七章　植原体病害防控管理 …… 172

一、监测预警 …… 172

二、农业防控 …… 172

三、物理防控 …… 173

四、化学防控 …… 173

五、生物防控 …… 174

六、生态防控 …… 175

参考文献 …… 175

第八章　展望 …… 177

一、检测诊断 …… 177

二、寄主范围 …… 178

三、植原体遗传特点及产生机制 …… 179

四、植原体-寄主互作机制研究 …… 180

五、绿色防控 …… 181

参考文献 …… 182

第一章

植原体研究概述

一、植原体生物学特征

植原体（phytoplasma）原称类菌原体（mycoplasma-like organism，MLO），1967年由日本学者Doi等首次发现（Doi et al.，1967）。1994年在第十届国际菌原体大会上正式将这一类微生物定名为"phytoplasma"。1997年，裘维蕃先生将"phytoplasma"翻译为"植原体"（IRPCM，2004；裘维蕃，1997）。植原体大小在50～1 000 nm，形态多样，包括球形、椭圆形、哑铃形、梭形、带状、长杆形等（刘仲健等，1999）。系统进化分析表明植原体为柔膜菌纲植原体候选属，与非固醇原体、螺原体等细菌的系统发育关系较近（Lim et al.，1989；Seemüller et al.，1998；Melamed et al.，2003）。植原体是引起植物病害的一类重要病原，寄主种类多，危害面积大，对社会经济发展和生态环境等造成严重影响（田国忠，1998，1999；田国忠等，2002；罗大全等，2008；蒯元璋，2012）。

植原体具有寄主依赖性，主要分布于植物的韧皮部和媒介昆虫的肠道、淋巴、唾液腺中。植原体能被抑制细菌蛋白质合成的四环素类抗生素所抑制，这是植原体与植物病毒最重要的一个区别。植原体主要通过刺吸式口器昆虫如叶蝉、飞虱、蟒等进行传播，也可以通过嫁接、菟丝子等进行传播（Hogenhout et al.，2008）。植原体侵染不同植物后可引起不同的症状，这主要与"植原体—寄主"间相互作用有关。随着植原体株系、植物种类、侵染时间、环境差异而变，而且植原体侵染引起的症状并不是单一的，而是综合表现出多种症状。植原体病害主要症状有丛枝、小叶、花器变态（绿变或花变叶）、叶片褪绿黄化、植株矮化、果实畸形等（宋传生，2011；陈旺，2015；于少帅等，2016）。

不同组植原体分布范围差异较大，16SrⅠ组翠菊黄化组植原体株系分布广，是多样性最丰富的一个16Sr组，可引起100余种植物病害（Lee et al.，2004）。而一些组的植原体分布地区则较窄，如引起水稻黄矮病（rice yellow dwarf disease）的16SrⅪ组植原体主要分布于亚洲。巴西报道的植原体病害主要由16SrⅠ和16SrⅢ组植原体引起（Mello et al.，

2006）。我国已报道的植原体主要属于16SrⅠ、16SrⅡ、16SrⅤ、16SrⅪⅩ和16SrⅩⅩⅫ组，其中16SrⅠ和Ⅴ组分布范围广、危害寄主多、造成的经济损失严重（赖帆等，2008）。植原体的地理分布和影响取决于其寄主范围和媒介昆虫的取食行为。某些寄主范围广、传毒媒介种类多的植原体如16SrⅠ组，分布范围也广，反之则分布范围窄（Lee et al., 2004；赖帆等，2008；Bertaccini & Duduk, 2009）。同组内不同16Sr亚组的植原体地理分布可能更加局限，如番茄巨芽病在不同国家和地区可由不同组植原体引起，如在美国由16SrⅠ组引起，在巴西由16SrⅢ组引起，在意大利由16SrⅠ、16SrⅤ和16SrⅪ组引起，而在约旦则由16SrⅥ组植原体引起（Mello et al., 2006）；16SrⅤ-B亚组植原体仅分布于亚洲的一些国家，而16SrⅤ-A亚组植原体则分布于欧洲和北美地区（Lee et al., 2004）；我国重阳木丛枝植原体为16SrⅤ-H亚组，目前该亚组仅在中国有报道（Lai et al., 2014）。

二、植原体基因组

1. 植原体全基因组

不同植原体基因组大小差异较大，根据脉冲场电泳测定，植原体基因组大小在530~1 350 kb（于少帅等，2016）。植原体基因组大小差异，会导致基因组在信息容量上的差异（Dickinson & Hodgetts, 2013）。基因组测序已完成的代表性植原体株系包括16SrⅠ组的洋葱黄化植原体（onion yellows phytoplasma, OY-M）（GenBank登录号AP006628）和翠菊黄化丛枝植原体（aster yellows witches'-broom phytoplasma, AYWB）（CP000061），16SrⅫ组的澳大利亚葡萄黄化植原体（*Candidatus* Phytoplasma australiense, PAa）（AM422018）和草莓致死黄化植原体（strawberry lethal yellows phytoplasma, SLY）（CP002548），16SrⅩ组的苹果簇生植原体（*Candidatus* phytoplasma mali, AT）（CU469464）等（Oshima et al., 2004；Bai et al., 2006；Tran-Nguyen et al., 2008；Kube et al., 2008；Andersen et al., 2013）。OY-M、AYWB、PAa和SLY染色体DNA为环状，而AT染色体DNA为线状。

不同植原体株系在基因结构和功能方面也存在差异，SLY含有1 126个开放阅读框（open reading frame, ORF），是目前全基因组已知的植原体中含有ORF最多的，而AT含有最少的ORF，仅有497个，PAa含有839个ORF，OY-M与AYWB全基因组中ORF数量相近，分别为754个和671个（Oshima et al., 2004；Bai et al., 2006；Tran-Nguyen et al., 2008；Kube et al., 2008；Andersen et al., 2013）。16SrⅠ组B亚组的OY-M比A亚组的AYWB植原体株系多154 kb碱基和83个ORF（Oshima et al., 2002；Oshima et al., 2004；Bai et al., 2006），造成这一差异的主要原因是AYWB比OY-M缺少总长达97 661 bp的多拷贝基因区域，而且这2个基因组中的基因簇的数量也明显不同。一些基因如Ⅰ型限制修饰系统的相关基因*hsdR*和*hsdM*，甘油酯代谢途径相关的基因*rfaG*等，在OY-M中存在，而在AYWB中则是缺失的。OY-M具有叶酸合成相关基因*folC*、*folK*和*folP*的全长序列，而AYWB则只具有剪截的*folK*和*folP*并且缺失*folC*。AYWB第423 992~660 824个碱基之间的序列与OY-M第354 087~103 752个碱基之间的序列，有大约250 kb的区域具有较高的相似性，这段区域表现为"同线性保守"。它包含了*tra5*插入序列之外的大部分代谢相关基因。在获得的3个洋葱黄化植原体株系中，完全测序的OY-M与另一个不能通过昆虫传播的植原体株系OY-NIM能引起丛枝症状，但不引起植株矮化症状；而野生型的OY-W既能引起丛枝也能引起矮化症状，并且诱导筛管增生和严重的筛管坏死，野生型OY-W植原体的基因组也比OY-M大了约130 kb，其中可能含有OY-M缺乏的与其致病性相关的基因（Oshima et al., 2001b）。

2. 植原体染色质外DNA

植原体染色质外DNA主要以质粒的形式存在，大小为3~11 kb，不同植原体中质粒在数量、大小、基因功能等方面均存在一定的差异（Dickinson & Hodgetts, 2013；于少帅等，2016）。植原体质粒最先在玉米丛矮植原体中发现（Davis et al., 1988），随后在其他植原体中也有发现（Nakashima & Hayashi, 1997；Liefting et al., 2004；Andersen et al., 2013）。翠菊黄化植原体AYWB中存在4个质粒，大小为3 872~5 104 bp（Bai et al., 2006），洋葱黄化植原体弱毒株系OY-M中存在2个质粒，与野生株系OY-W和非昆虫传播株系OY-NIM质粒大小和数目均不相同，其中OY-NIM中的质粒pOYNIM与OY-M的质粒pOYM相比，缺少2个阅读框ORF3和ORF4，ORF3编码的细胞膜蛋白可能在细胞膜的界面上与介体昆虫的互作中发挥作用（Nishigawa et al., 2002a；Ishii et al., 2009a, b）。草莓致死黄化植原体SLY株系中存在1个长3 635 bp的质粒pPASb11（Andersen et al., 2013），澳大利亚葡萄黄化植原体PAa存在1个长3.7 kb的质粒（Tran-Nguyen et al., 2008），翠菊黄化植原体AYWB中存在4个质粒，大小为3 872~5 104 bp（Bai et al., 2006）。

不同质粒中ORF的数目及功能也存在一定差异。已知植原体质粒的ORF数目为2~11个，pBLTVA-1质粒具有11个ORF，而同一种植原体内的另一个质粒pBLTVA-2则只有2个ORF（Liefting et al., 2004）。国内先后鉴定了几种16SrⅠ组植原体的质粒，结果表明，即使同组内的植原体质粒特性差异也相当明显。在泡桐丛枝植原体中鉴定了2个质粒，质粒pPaWBNy-1长4 485 bp，推测含有6个ORF，质粒pPaWBNy-2长3 837 bp，含有5个ORF（Lin et al., 2009）。在苦楝丛枝植原体中鉴定了1个质粒pCWBFq，长4 446 bp，可编码6种蛋白（宋传生等，2011）。在桑萎缩、长春花绿变、泡桐丛枝和苦楝丛枝植原体中均检测到质粒的存在，大小为3 833~3 943 bp，4种质粒皆编码5种蛋白（胡佳续等，2013）。

质粒的进化速度比染色体快得多，这种现象在植原体遗传变异发生过程中扮演着重要角色，并可能赋予植原体迅速适应环境的能力，或造成植原体致病性的差异（于少帅等，2016）。在pBLTVA-1中存在着很多重复序列，它们参与质粒DNA重组，pBLTVA-2被认为来源于pBLTVA-1，是pBLTVA-1基因重组的结果（Liefting et al., 2004）。OY-W株系质粒

EcOYW2的ORF5（rep）和ORF4（ssb）之间的非编码区，与EcOYW1上游序列及pOYW下游序列相似，表明EcOYW1和pOYW之间的这段共同序列可能是重组点，2个质粒通过这个重组点产生了EcOYW2，这种质粒间的重组过程在细菌遗传多样性产生和提供对新环境的迅速适应能力方面发挥重要的进化作用（Nishigawa et al.，2002b）。重组不仅发生在质粒之间，也可发生在质粒和染色体之间甚至质粒和寄主之间（Melamed et al.，2003）。植物病原菌的质粒在致病性和毒性方面起着重要的作用（Liefting et al.，2004），质粒多样性使得植原体在致病性和被介体昆虫传播方面也表现出多样性特征，这可能与染色质外基因重组、启动子的缺失和突变、不同植原体中质粒的数量、不同寄主中同一质粒的拷贝数、转录和翻译的数量等有关（于少帅等，2016）。

3. 植原体编码基因

植原体一般含有2个rRNA操纵子，rRNA操纵子包括16S rDNA、23S rDNA、16S—23S rDNA间区，其组织模式是5′-16S-23S-3′。16S rRNA基因是植原体中的高度保守，但在不同种类植原体间又存在明显变异的序列，植原体之间相似性为88%～99%，与亲缘关系最近的非固醇原体相似性为87%～88.5%（Seemüller et al.，1998）。所以在植原体分子分类和鉴定的研究中，16S rRNA基因是应用最广的遗传标记，也揭示了植原体丰富的遗传多样性特征。但是在某些种类植原体中2个rRNA操纵子存在着序列的变异，比如新西兰亚麻黄叶植原体2个操纵子的16S rRNA基因序列的差异达0.27%（Liefting et al.，1996），而巴西番茄巨芽植原体16S rRNA基因序列有3种不同的类型（Mello et al.，2006）。23S rRNA基因也已用于植原体系统进化和分类鉴定研究中，提高了植原体的鉴别能力（Hodgetts et al.，2008；Lai et al.，2014）。

核糖体蛋白操纵子基因（*rp*）也是反映植原体遗传多样性特征的基因（Lim & Sears，1992）。*rp*基因与16S rRNA基因相比，变异性较大。在16S rRNA基因对植原体进行分组的基础上，*rp*基因可以在亚组水平上对植原体进一步分类，从而更充分地揭示植原体的遗传多样性（Lee et al.，1998）。基于*rp*序列建立的进化树可揭示出植原体之间更为细微的遗传进化关系，在一定程度上揭示了植原体不同株系间更加丰富的遗传信息（Lee et al.，1998；Martini et al.，2007）。植原体编码延伸因子基因*tuf*和*fusA*分别编码蛋白延伸因子EF-Tu和EF-G，其中*tuf*基因序列和16S rRNA基因具有相对低的保守性（Lee et al.，2000），可在16S rRNA基因分组的基础上用于亚组的补充区分（Schneider et al.，1997；Koui et al.，2003；Yu et al.，2014）。Streten和Gibb（2005）利用*tuf*基因序列将16SrXII-B中的*Ca.* P. australiense株系细分为16SrXII-B（*tuf*-Australia I）、16SrXII-B（*tuf*-New Zealand I）、16SrXII-B（*tuf*-New Zealand II）。运蛋白编码基因*secA*和*secY*比16S rRNA、*rp*、*tuf*和*fusA*等基因序列有更大的变异，可以在植原体组或候选种以下的鉴别和遗传多样性研究上有更大的价值（Lee et al.，2006；Hodgetts et al.，2008；Yu et al.，2014）。胸苷酸激酶基因*tmk*的产物是三磷酸胸苷从头合成和补救途径的关键酶，广泛存在于各种生物中，通过 *tmk* 基因分析揭示了我国枣疯和泡桐丛枝植原体存在不同的 *tmk* 基因型和丰富的遗传多样性。从泡桐丛枝植原体平山株系 PaWBPs 和吉安株系 PaWBJan 中各克隆测序的 93 条和 41 条 *tmk-a* 中，分别有 52 种和 24 种不同的序列，*tmk-a* 的 ORF 可被划分为 2 类，*tmk-a-1* 为 639 bp 和 *tmk-a-2* 为 627 bp，PaWBPs 株系 *tmk-a-1* 与 *tmk-a-2* 序列条数的比值约为 2.5，而 PaWBJan 株系为 3.3，PaWBPs 有 5 个相同的 *tmk-a-1* 序列与 PaWBJan 的 1 个 *tmk-a-1* 序列及枣疯植原体 *tmk-Y* 序列完全一致；但 *tmk-b* 可能是单拷贝基因，且与 16S rRNA 等保守基因的进化趋势更一致（徐启聪等，2009；宋传生，2014）。

4. 植原体非编码区

植原体非编码区包括启动子、16S—23S rRNA 基因间隔区、*fusA-tuf* 基因间隔区等。启动子是调控编码基因转录的非编码区。Palmano 等（2001）首次将西方 X 病植原体 16S rRNA 基因上游的启动子序列构建到无启动子的报道基因 *cat* 的质粒上，研究证明其能在芽孢杆菌中启动基因的表达。Ishii 等（2009a）用启动子软件对植原体质粒序列上的启动子进行了预测和分析，推断 ORF3 编码蛋白在变异株系 OY-NIM 中不表达可能是由于启动子的缺失或突变造成的，而且质粒和染色体 DNA 上不同基因的启动子在−35 区和−10 区的特征序列核苷酸都有差异（Jung et al.，2003）。于少帅等（2016）在 16SrⅠ组泡桐丛枝植原体、苦楝丛枝植原体、莴苣黄化植原体 *tuf* 基因上游长 129 bp 的区域预测有完整的启动子保守结构并对其启动子活性进行了鉴定。植原体 16S—23S rRNA 与 *fusA—tuf* 基因间隔区，受到的进化阻力比较小，在遗传上较 16S rRNA 基因有更大的可变性。因此，依据 16S rRNA 和 23S rRNA 之间的间隔区域、*fusA* 和 *tuf* 之间的基因间隔区域或 tRNA 侧翼的可变区域建立的植原体分类体系与依据 16S rRNA 基因建立的体系基本一致（Wang & Hiruki，2005；于少帅等，2016）。Wang 和 Hiruki（2005）通过对植原体 16S—23S rRNA 基因间区序列 PCR 产物的异源双链迁移率分析，将 62 个植原体株系在亚组水平上进行了分类。于少帅等（2016）基于 *fusA* 和 *tuf* 基因间区序列，对泡桐丛枝植原体（PaWB）、苦楝丛枝（CWB）、莴苣黄化（LY）、桑萎缩（MD）、长春花绿变（PeV）等 16SrⅠ组和枣疯病（JWB）、樱桃致死黄化（CLY）、重阳木丛枝（BiWB）、黄金槐丛枝（SoWB）等 16SrⅤ组株系构建的系统树可清晰区分不同组别的植原体株系。

5. 植原体操纵子

操纵子学说最早由法国巴斯德研究所著名科学家 Jacob 和 Monod 于 1961 年提出，典型的操纵子模型结构包括结构基因、调控元件和调节基因 3 个部分。操纵子结构基因发生变异只影响其编码的特定蛋白氨基酸序列及结构，而调控基因序列的变异则可能影响整个操纵子所有结构基因的表达（李明刚，2004）。目前，对操纵子基因结构和表达调控机制了解相对较多的操纵子有大肠杆菌乳糖操纵子（lactose operon，*lac*）、色氨酸操纵子（tryptophan operon，*trp*）、半乳糖操纵子（galactose operon，*gal*）、阿拉伯糖操纵子（arabinose operon, *ara*）、rRNA操纵子（*rrnE*）等（朱玉贤等, 2002；特纳等, 2010）。大肠杆菌的蛋白延伸因子Tu编码基因*tuf*与核糖体蛋白编码基因*rps12*和*rps7*、蛋白延伸因子G编码基因*fusA*组成一个*str*操纵子，其基因结构排列顺序为5′-*rps12-rps7-fusA-tuf*-3′，*rps12*基因邻接上游序列和*tuf*基因邻接下游序列存在启动子和转录终止结构（Post et al., 1980）。Sanangelantoni等（1993）研究发现*Methanococcus vannielii*、Cyanelles等细菌的核糖体蛋白编码基因*rpS10*与*tuf*基因连在一起，与*str*操纵子一起转录。而在大肠杆菌中*rpS10*基因在*str*操纵子下游15 kb处，与核糖体蛋白编码基因*rpl3*、*rpl4*等组成*S10*操纵子，在*S10*操纵子3′端发现转录终止结构（Post et al., 1980；Zurawski et al., 1985）。*S10*操纵子和*spc*操纵子在大肠杆菌基因组中均有发现（Post et al., 1980；Zurawski et al., 1985）。Miyata等（2002a）研究发现洋葱黄化植原体*S10*操纵子和*spc*操纵子之间没有找到与转录起始或终止相关的序列，在*S10*操纵子上游邻接序列发现了可能的启动子结构，推测*S10*操纵子和*spc*操纵子在植原体中是一个转录单位，构成*S10-spc*操纵子。Miyata等（2002b）研究发现洋葱黄化植原体OY的*str*操纵子的结构为5′-*rps12-rps7-fusA-tuf*-3′，与大肠杆菌的*str*操纵子结构排布一致。与支原体相比，洋葱黄化植原体OY的*str*操纵子基因结构排布与芽孢杆菌更接近。于少帅等（2018）扩增了我国泡桐丛枝植原体、莴苣黄化植原体株系*tuf*基因及其上游6个基因*rplL*、*rpoB*、*rpoC*、*rps12*、*rps7*、*fusA*序列，比较分析发现泡桐丛枝植原体PaWB-sdyz和PaWB-fjfz、莴苣黄化植原体LY-fjya1、洋葱黄化植原体OY-M、翠菊黄化植原体AYWB、澳大利亚葡萄黄化植原体PAa、草莓致死黄化植原体SLY、苹果簇生植原体AT等不同组的植原体株系*tuf*与其上游6个基因的结构顺序皆为序列5′-*rplL-rpoB-rpoC-rps12-rps7-fusA-tuf*-3′，结合基因表达初步分析发现相关基因间可能也存在一个操纵子结构。

三、我国热区植原体遗传多样性

植原体因其遗传多样性丰富，在流行监测与防控管理方面带来了诸多挑战。由植原体引起的植物病害是相关作物种植上面临的重大病害防控难题，该类病害对社会经济和生态环境影响严重。我国热区生物多样性丰富，植原体种类也较为丰富且相对独特。我国热区已报道植原体病害120余种，占我国已报道植原体病害的60%以上。海南是我国特色的热带岛屿省份，其植原体病害十分丰富，约占我国植原体病害的1/5，植原体病害分布地区几乎遍布整个海南岛（Che et al., 2024）。16SrXXXII组植原体与槟榔、椰子、油棕等多种热带经济棕榈作物致死性病害相关（Nejat et al., 2009, 2013；Yu et al., 2023）。目前，16SrXXXII组植原体仅在我国海南、广东、云南等热区省份有报道，可侵染槟榔、柚子、喜树、山黄麻等我国热区经济作物或绿化植物（王柱华等, 2021；席亦民等, 2022；Yu et al., 2023）。植原体可侵染的植物寄主种类十分丰富，因此对我国热区相关植物的种植业与产业造成了严重的影响。如由植原体引起的槟榔黄化病，给海南槟榔种植及相关产业造成毁灭性打击（车海彦等，2010；于少帅等，2021）。由植原体引起的甘蔗白叶病，给我国热区甘蔗种植及相关产业带来严重影响（李文凤等，2021）。国外已报道的一些植原体病害在我国热区虽未见报道，但极具入侵风险，如我国重大检疫性植原体病害椰子致死性黄化病等，椰子是我国海南主要的特色作物，具有重要的经济、绿化和生态价值（卢丽兰等，2021）。截至目前，椰子致死性植原体在国内尚未见报道，但海南是我国椰子主产区，椰子致死性植原体在中国的适应性分析和入侵中国的风险分析表明，我国为椰子植原体的适生区，特别是华南沿海及海南大部分地区为中、高风险区（朱辉等，2010；曹学仁等，2014）。因此，明确我国热区植原体种类及其病害的多样性及特征，是对广布于我国热区的这类致死性病害实施精准监测和有效防控的前提。

参考文献

曹学仁，车海彦，杨毅，等，2014. 基于Maxent的椰子致死性黄化植原体在中国的适生性分析[J]. 热带作物学报，35（11）：2260-2265.

车海彦，吴翠婷，符瑞益，等，2010. 海南槟榔黄化病病原物的分子鉴定[J]. 热带作物学报，31（1）：83-87.

陈旺，2015. 小麦蓝矮植原体基因组序列测定及其分泌蛋白功能研究[D]. 杨凌：西北农林科技大学.

胡佳续，宋传生，林彩丽，等，2013. 四种植物病害植原体病原质粒全序列测定及分子特征[J]. 林业科学，49（4）：90-97.

蒯元璋，2012. 桑树病原原核生物及其病害的研究进展（Ⅱ）[J]. 蚕业科学，38（5）：898-913.

赖帆，李永，徐启聪，等，2018. 植原体的最新分类研究动态[J]. 微生物学通报，35（2）：291-295.

李明刚，2004. 高级分子遗传学[M]. 北京：科学出版社.

李文凤，王晓燕，仓晓燕，等，2021. 甘蔗白叶病抗病性鉴定方法的建立与应用[J]. 植物保护，47（5）：245-248.

刘仲健，罗焕亮，张景宁，1999. 植原体病理学[M]. 北京：中国林业出版社.

卢丽兰，刘蕊，肖勇，等，2021. 椰子种质资源、栽培与利用研究进展[J]. 热带作物学报，42（6）：1795-1803.

罗大全，车海彦，刘先宝，等，2008. 海南苦楝丛枝病植原体的分子鉴定[J]. 热带作物学报，29（4）：522-524.

裘维蕃，1997. 关于在植物病理学及相关学科中一些术语翻译的商榷[J]. 植物病理学报，27

（2）：104-106.

宋传生，2011. 木本植物黄化类植原体的检测鉴定及苦楝丛枝植原体质粒的测定[D]. 北京：中国林业科学研究院.

宋传生，林彩丽，田国忠，等，2011. 苦楝丛枝植原体质粒的测定与分子特征[J]. 微生物学报，51（9），1158-1167.

特纳，2010. 分子生物学：第3版[M]. 刘进元，刘文颖，译. 北京：科学出版社.

田国忠，1998. 枣疯病的预防和治疗策略研究[J]. 林业科技通讯，（2）：14-16.

田国忠，1999. 北京地区木本植物植原体病害发生及防治对策[J]. 北京农业科学，17（6）：1-5.

田国忠，张志善，李志清，等，2002. 我国不同地区枣疯病发生动态和主导因子分析[J]. 林业科学，38（2）：83-91.

王柱华，王文鹏，袁思平，等，2021. 喜树丛枝植原体的分子鉴定及Taq Man探针实时荧光定量PCR检测方法的建立[J]. 植物病理学报，51（3）：429-440.

席亦民，2022. 半寄生植物广寄生侵染过程初探及其丛枝病植原体鉴定[D]. 广州：中国林业科学研究院热带林业研究所.

徐启聪，田国忠，王振亮，等，2009. 中国各地不同枣树品种上枣疯病植原体的PCR检测及分子变异分析[J]. 微生物学报，49（11）：1510-1519.

于少帅，林彩丽，潘皎，等，2016. 泡桐丛枝和枣疯病植原体tuf基因上游序列结构、功能和遗传变异比较分析[J]. 微生物学通报，43（5）：1060-1069.

于少帅，林彩丽，王圣洁，等，2018. 植原体tuf基因与其上游部分基因结构和相关基因启动子保守区域特征及活性分析[J]. 生物多样性，26（7）：738-748.

于少帅，宋薇薇，覃伟权，2021. 海南槟榔黄化植原体分子检测及其系统发育关系研究[J]. 热带作物学报，42（11）：3066-3072.

于少帅，徐启聪，林彩丽，等，2016. 植原体遗传多样性研究现状与展望[J]. 生物多样性，24（2）：205-215.

朱辉，覃伟权，吴多扬，等，2010. 椰子致死性黄化植原体传入中国的风险性分析[J]. 江西农业学报，22（11）：84-87.

朱玉贤，李毅，2002. 现代分子生物学[M]. 2版. 北京：高等教育出版社.

ANDERSEN M T, LIEFTING L W, HAVUKKALA I, et al., 2013. Comparison of the complete genome sequence of two closely related isolates of 'Candidatus phytoplasma australiense' reveals genome plasticity[J]. BMC Genomics, 14: 529.

BAI X D, ZHANG J H, EWING A, et al., 2006. Living with genome instability: the adaptation of phytoplasma to diverse environments of their insect and plant hosts[J]. Journal of Bacteriology, 188: 3682-3696.

BERTACCINI A, DUDUK B, 2009. Phytoplasma and phytoplasma diseases: a review of recent research[J]. Phytopathologia Mediterranea, 48: 355-378.

CHE H Y, YU S S, CHEN W, et al., 2024. Molecular identification and characterization of novel taxonomic subgroups and new host plants in 16SrⅠand 16SrⅡ group phytoplasmas and their evolutionary diversity on Hainan Island, China[J]. Plant Disease, doi: PDIS-12-23-2682-RE.

DICKINSON M, HODGETTS J, 2013. Phytoplasma: methods and protocols[M]. Totoma: Humana Press.

DOI Y, TERANAKA M, YORA K, et al., 1967. Mycoplasma or PLT group like microorganisms found in the phloem elements of plants infected with mulberry dwarf, potato witches'-broom, aster yellows or pauwlonia witches'-broom[J]. Annals of the Phytopathological Society of Japan, 33: 259-266.

HODGETTS J, BOONHAM N, MUMFORD R, et al., 2008. Phytoplasma phylogenetics based on analysis of *secA* and 23S rRNA gene sequences for improved resolution of candidate species of '*Candidatus* Phytoplasma' [J]. International Journal of Systematic and Evolutionary Microbiology, 58: 1826-1837.

HOGENHOUT S A, OSHIMA K, ELD A, et al., 2008. Phytoplasmas: bacteria that manipulate plants and insects[J]. Molecular Plant Pathology, 9 (4): 403-423.

IRPCM, 2004. '*Candidatus* Phytoplasma', a taxon for the wall-less, non-helical prokaryotes that colonize plant phloem and insects[J]. International Journal of Systematic and Evolutionary Microbiology, 54: 1243-1255.

ISHII Y, KAKIZAWA S, HOSHI A, et al., 2009a. In the non-insect-transmissible line of onion yellows phytoplasma (OY-NIM), the plasmid-encoded transmembrane protein ORF3 lacks the major promoter region[J]. Microbiology, 155: 2058-2067.

ISHII Y, OSHIMA K, KAKIZAWA S, et al., 2009b. Process of reductive evolution during 10 years in plasmids of a non-insect-transmissible phytoplasma[J]. Gene, 446: 51-57.

JUNG H Y, MIYATA S, OSHIMA K, et al., 2003. First complete nucleotide sequence and heterologous gene organization of the two rRNA operons in the phytoplasma genome[J]. DNA and Cell Biology, 22: 209-215.

KOUI T, NATSUAKI T, OKUDA S, 2003. Phylogenetic analysis of elongation factor Tu gene of phytoplasmas from Japan[J]. Journal of General Plant Pathology, 69: 316-319.

KUBE M, SCHNEIDER B, KUHL H, et al., 2008. The linear chromosome of the plant-pathogenic mycoplasma '*Candidatus* phytoplasma mali' [J]. BMC Genomics, 9: 306.

LAI F, SONG C S, REN Z G, et al., 2014. Molecular characterization of a new member of the 16Sr Ⅴ group of phytoplasma associated with *Bischofia polycarpa*（Levl.）Airy Shaw witches'-broom disease in China by a multiple gene-based analysis[J]. Australian Plant Pathology, 43: 557-569.

LEE I M, DAVIS R E, GUNDERSEN-RINDAL D E, 2000. Phytoplasma: Phytopathogenic Mollicutes[J]. Annual Review of Microbiology, 54: 221-255.

LEE I M, GUNDERSEN-RINDAL D E, DAVIS R E, et al., 1998. Revised classification scheme of phytoplasmas based on RFLP analyses of 16S rRNA and ribosomal protein gene sequence[J]. International Journal of Systematic Bacteriology, 48: 1153-1169.

LEE I M, GUNDERSEN-RINDAL D E, DAVIS R E, et al., 2004. '*Candidatus* Phytoplasma asteris', a novel phytoplasma taxon associated with aster yellows and related diseases[J]. International Journal of Systematic and Evolution Microbiology, 54: 1037-1048.

LEE I M, ZHAO Y, BOTTNER K D, 2006. *SecY* gene sequence analysis for finer differentiation of diverse strains in the aster yellows phytoplasma group[J]. Molecular and Cellular Probes, 20: 87-91.

LIEFTING L W, ANDERSEN M T, BEEVER R E, et al., 1996. Sequence heterogeneity in two 16S rRNA genes of Phormium yellow leaf phytoplasma[J]. Applied and Environmental Microbiology, 62: 3133-3139.

LIEFTING L W, SHAW M E, KIRKPATRIC B C, 2004. Sequence analysis of two plasmids from the phytoplasma beet leafhopper-transmitted virescence agent[J]. Microbiology, 150: 1809-1817.

LIM P O, SEARS B B, 1989. 16S rRNA sequence indicates that plant-pathogenic mycoplasma like organisms are evolutionarily distinct from animal mycoplasmas[J]. Journal of Bacteriology, 171（11）: 5901-5906.

LIM P O, SEARS B B, 1992. Evolutionary relationships of a plant-pathogenic mycoplasmalike organism and *Acholeplasma laidlawii* deduced from two ribosomal protein gene sequences[J]. Journal of Bacteriology, 174: 2606-2611.

LIN C L, ZHOU T, LI H F, et al., 2009. Molecular characterization of two plasmids from paulownia witches'-broom phytoplasma and detection of a plasmid-encoded protein in infected plants[J]. European Journal of Plant Pathology, 123: 321-330.

MARTINI M, LEE I M, BOTTNER K D, et al., 2007. Ribosomal protein gene-based phylogeny for finer differentiation and classification of phytoplasmas[J]. International Journal of Systematic and Evolutionary Microbiology, 57: 2037-2051.

MELAMED S, TANNE E, BEN-HAIM R, et al., 2003. Identification and characterization of phytoplasmal genes, employing a novel method of isolating phytoplasmal genomic DNA[J].

Journal of Bacteriology, 185（22）：6513-6520.

MELLO A P O A, BEDENDO I P, CAMARGO L E A, 2006. Sequence heterogeneity in the 16S rDNA of tomato big bud phytoplasma belonging to group 16SrⅢ[J]. Journal of Phytopathology, 154：245-249.

MIYATA S, FURUKI K, OSHIMA K, et al., 2002a. Complete nucleotide sequence of the *S10-spc* operon of phytoplasma: gene organization and genetic code resemble those of *Bacillus subtilis*[J]. DNA and Cell Biology, 21（7）：527-534.

MIYATA S, FURUKI K, SAWAYANAGI T, et al., 2002b. Gene arrangement and sequence of *str* operon of phytoplasma resemble those of *Bacillus* more than those of *Mycoplasma*[J]. Journal of General Plant Pathology, 68（68）：62-67.

NAKASHIMA K, HAYASHI T, 1997. Sequence analysis of extrachromosomal DNA of sugarcane white leaf phytoplasma[J]. Annals of the Phytopathological Society of Japan, 63：21-25.

NEJAT N, SIJAM K, ABDULLAH S N A, et al., 2009. Phytoplasmas associated with disease of coconut in Malaysia: phylogenetic groups and host plant species[J]. Plant Pathol. 58：1152-1160.

NEJAT N, VADAMALAI G, DAVIS R E, et al., 2013. 'Candidatus Phytoplasma malaysianum', a novel taxon associated with virescence and phyllody of Madagascar periwinkle (*Catharanthus roseus*) [J]. International Journal of Systematic and Evolutionary Microbiology, 63：540-548.

NISHIGAWA H, OSHIMA K, KAKIZAWA S, et al., 2002a. plasmid from a non-insect-transmissible line of a phytoplasma lacks two open reading frames that exist in the plasmid from the wild-type line[J]. Gene, 298：195-201.

NISHIGAWA H, OSHIMA K, KAKIZAWA S, et al., 2002b. Evidence of intermolecular recombination between extrachromosomal DNAs in phytoplasma: a trigger for the biological diversity of phytoplasm[J]. Microbiology, 148：1389-1396.

OSHIMA K, KAKIZAWA S, NISHIGAWA H, et al., 2004. Reductive evolution suggested from the complete genome sequence of a plant-pathogenic phytoplasma[J]. Nature Genetics, 36：27-29.

OSHIMA K, MIYATA S, SAWAYANAGI T, et al., 2002. Minimal set of metabolic pathways suggested from the genome of onion yellows phytoplasma[J]. Journal of General Plant Pathology, 68：225-236.

OSHIMA K, SHIOMI T, KUBOYAMA T, et al., 2001b. Isolation and characterization of derivative lines of the onion yellows phytoplasma that do not cause stunting or phloem hyperplasia[J]. Phytopathology, 91: 1024-1029.

PALMANO S, KIRKPATRICK B C, FIRRAO G, 2001. Expression of chloramphenicol acetyltransferase in *Bacillus subtilis* under the control of a phytoplasma promoter[J]. FEMS Microbiology Letters, 199: 177-179.

POST L E, NOMURA M, 1980. DNA sequences from the *str* operon of *Escherichia coli*[J]. The Journal of Biological Chemistry, 255 (10): 4660-4666.

SANANGELANTONI A M, TIBONI O, 1993. The chromosomal location of genes for elongation factor Tu and ribosomal protein *S10* in the cyanobacterium Spirulina platensis provides clues to the ancestral organization of the *str* and *S10* operons in prokaryotes[J]. Journal of General Microbiology, 139 (11): 2579-2584.

SCHNEIDER B, GIBB K S, SEEMÜLLER E, 1997. Sequence and RFLP analysis of the elongation factor Tu gene used in differentiation and classification of phytoplasmas[J]. Microbiology, 143: 3381-3389.

SEEMÜLLER E, MARCONE C, LAUER U, et al., 1998. Current status of molecular classification of the phytoplasma[J]. Journal of Plant Pathology, 80 (1): 3-26.

STRETEN C, GIBB K S, 2005. Genetic variation in *Candidatus* Phytoplasma australiense[J]. Plant Pathology, 54: 8-14.

TRAN-NGUYEN L T T, KUBE M, SCHNEIDER B, et al., 2008. Comparative genome analysis of 'Candidatus phytoplasma australiense' (subgroup *tuf*-Australia I ; *rp*-A) and 'Ca. phytoplasma asteris' strains OY-M and AY-WB[J]. Journal of Bacteriology, 190: 3979-3991.

WANG K, HIRUKI C, 2005. Distinctions between phytoplasmas at the subgroup level detected by heteroduplex mobility assay[J]. Plant Pathology, 54: 625-633.

YU S S, LI Y, REN Z G, et al., 2014. Multilocus sequences confirm genetic variation and differentiation among 16Sr I group phytoplasma strains infecting environmentally and economically important plants in different regions of China[C]//Proceedings of the Annual Meeting of Chinese Society for Plant Pathology. Bejing: China Agricultural Science and Technology Press: 366-367.

YU S S, ZHU A N, CHE H Y, et al., 2023. Molecular identification of 'Candidatus Phytoplasma malaysianum' -related strains associated with *Areca catechu* palm yellow leaf disease and phylogenetic diversity of the phytoplasmas within 16Sr XXXII Group[J]. Plant Disease, doi: PDIS-11-23-2275-RE.

ZURAWSKI G, ZURAWSKI S M, 1985. Structure of the *Escherichia coli S10* ribosomal protein operon[J]. Nucleic Acids Research, 13 (12): 4521-4526.

第二章

植原体系统分类

植原体是寄生于植物韧皮部、无细胞壁、尚不能人工分离培养的一类重要原核致病菌（Doi et al., 1967; Dickinson & Hodgetts, 2013）。长期以来，由于植原体离体培养困难，研究进展较慢。近20年来，随着分子技术的广泛应用，已经充分明确了植原体的系统进化和分类地位，即属于细菌界（Bacteria）、软壁菌门（Tenericutes）、柔膜菌纲（Mollicutes）、无胆甾原体目（Acholeplasmatales）、无胆甾原体科（Acholeplasmataceae）、植原体候选属（Candidatus Phytoplasma）。在柔膜菌纲内，植原体形成一个大的独立的进化分枝，与柔膜菌纲内的无胆甾原体属（Acholeplasma）和螺原体属（Spiroplasma）的系统进化关系较近，在柔膜菌纲以外，植原体与GC含量较低的革兰氏阳性菌芽孢杆菌属（Bacillus）、梭菌属（Clostridium）、链球菌属（Streptococcus）等的系统进化关系较近（Lim et al., 1989; Melamed et al., 2003）。

一、植原体16Sr组

在植原体遗传变异和系统发育研究中应用较多的保守基因有16S rDNA、rp、tuf、secA、secY、fusA、rpoB等（蔡红等，2003; Li et al., 2014; Foissac et al., 2013; Valiunas et al., 2013）。16S rDNA在植原体中有2个重复，且具有一定的差异（Liefting et al., 1996; Mello et al., 2006）。16S rDNA序列是原核生物较为理想的分子进化标签，因此基于16S rDNA的植原体分类系统最先建立。rp、tuf等基因与16S rRNA基因相比，变异性较大，可作为植原体基于16S rRNA基因分组的补充，在一定程度上揭示植原体株系间更加丰富、细腻的多样性（Lim & Sears, 1992; Lee et al., 1998; Martini et al., 2007; Lee et al., 2000; Koui et al., 2003; Schneider et al., 1997）。

基于16S rDNA序列分析，Lee等（1998）将34个植原体株系分成14个组，Seemüller等（1998）将57个株系分为20个植原体组。Wei等（2007）基于计算机模拟的RFLP分析建立了植原体在线分类鉴定系统iPhyClassifier，并通过对已发布的植原体16S rRNA基因序列进行模拟RFLP分析，将800多个植原体序列分为28个组。目前，共37个植原体16Sr组被报道，如表2-1所示。

二、植原体候选种

'Candidatus'一词于1994年首次用于描述难以培养的细菌，根据16S rRNA基因序列赋予潜在分类群适当的地位（Murray et al.，1994）。1995年，基于植原体16S rRNA基因序列，第一个植原体候选种（'Candidatus Phytoplasma'）'Ca. Phytoplasma orantifolia'被命名（Zreik et al.，1995）。2004年，植原体候选种命名规则由国际比较支原体学研究计划署（International Research Programme on Comparative Mycoplasmology，IRPCM）植原体/螺旋体工作组-植原体分类学组提出（IRPCM，2004）。2022年，学者对植原体候选种命名规则进行修订（Bertaccini et al.，2022）。目前，共49个植原体候选种被报道，如表2-1所示。

2019年，Bergey's手册建议细菌分类应由基于16S rRNA基因序列水平向基于基因组水平转变（Wei & Zhao，2022；Hugenholtz et al.，2021）。2004年首个植原体（洋葱黄化植原体OY-M株系）基因组测序完成（Oshima et al.，2004）。已有属于13个组和29个亚组的植原体株系共47个植原体株系的基因组测序完成，包括12个全基因组和35个基因组草图，这些植原体基因组信息如表2-2所示（Wei & Zhao，2022）。

植原体候选种系统分类及新种的鉴定，既有基于植原体16S rRNA基因序列水平的，也有基于植原体基因组水平的（Wei & Zhao，2022；Bertaccini et al.，2022）。目前，植原体已有39个组150多个亚组1 000余种株系，但只有小部分植原体株系基因组被测序完成（Wei & Zhao，2022；Bertaccini et al.，2022）。植原体基因组数据将揭示植原体更为丰富的遗传信息，为植原体的系统分类提供更好的、更全面的依据。

表2-1 植原体16Sr组/亚组与植原体候选种（'*Ca.* Phytoplasma'）信息表（Wei & Zhao, 2022）

组	植原体候选种数量	植原体候选种名称	参考株系序列号	亚组	参考文献
16SrⅠ: Aster yellows group	3	'*Ca.* Phytoplasma asteris'	M30790	16SrⅠ-B	Lee et al., 2004a
		'*Ca.* Phytoplasma lycopersici'	EF199549	16SrⅠ-Y	Arocha et al., 2007
		'*Ca.* Phytoplasma tritici'	NZ AVAO01000003	16SrⅠ-C	Zhao et al., 2021
16SrⅡ: Peanut witches'-broom group	1	'*Ca.* Phytoplasma aurantifolia'	U15442	16SrⅡ-B	Zreik et al., 1995
	*Abolished	'*Ca.* Phytoplasma australasia'	Y10096	16SrⅡ-D	White et al., 1998
16SrⅢ: X-disease group	1	'*Ca.* Phytoplasma pruni'	JQ044393	16SrⅢ-A	Davis et al., 2013
16SrⅣ: Coconut lethal yellows group	2	'*Ca.* Phytoplasma palmae'	U18747	16SrⅣ-A	IRPCM, 2004; Bertaccini et al., 2022
		'*Ca.* Phytoplasma cocostanzaniae'	X80117	16SrⅣ-C	IRPCM, 2004; Bertaccini et al., 2022
16SrⅤ: Elm yellows group	4	'*Ca.* Phytoplasma ulmi'	AY197655	16SrⅤ-A	Lee et al., 2004b
		'*Ca.* Phytoplasma ziziphi'	AB052876	16SrⅤ-B	Jung et al., 2003
		'*Ca.* Phytoplasma rubi'	AY197648	16SrⅤ-E	Malembic-Maher et al., 2011
		'*Ca.* Phytoplasma balanitae'	AB689678	16SrⅤ-new subgroup	Win et al., 2013
16SrⅥ: Clover proliferation group	2	'*Ca.* Phytoplasma trifolii'	AY390261	16SrⅥ-A	Hiruki et al., 2004
		'*Ca.* Phytoplasma sudamericanum'	GU292081	16SrⅥ-I	Davis et al., 2012
16SrⅦ: Ash yellows group	1	'*Ca.* Phytoplasma fraxini'	AF092209	16SrⅦ-A	Griffiths et al., 1999

（续表）

组	植原体候选种数量	植原体候选种名称	参考株系序列号	亚组	参考文献
16Sr XXIX：Cassia witches'-broom group	1	'Ca. Phytoplasma omanense'	EF666051	16Sr XXIX-A	Al-Saady et al., 2008
16Sr XXX：Salt cedar witches'-broom group	1	'Ca. Phytoplasma tamaricis'	FJ432664	16Sr XXX-A	Zhao et al., 2009
16Sr XXXI：Soybean stunt phytoplasma group	1	'Ca. Phytoplasma costaricanum'	HQ225630	16Sr XXXI-A	Lee et al., 2011
16Sr XXXII：Malaysian periwinkle virescence group	1	'Ca. Phytoplasma malaysianum'	EU371934	16Sr XXXII-A	Nejat et al., 2013
16Sr XXXIII：Allocasuarina group	1	'Ca. Phytoplasma allocasuarinae'	AY135523	16Sr XXXIII-A	IRPCM, 2004
16Sr XXXIV：grapevine yellows		No new species identified, abolished	DQ232752		
16Sr XXXV：Pepper witches'-broom group		No new species identified, abolished	EU125184		
16Sr XXXVI：foxtail palm yellow decline group	1	'Ca. Phytoplasma wodyetiae'	KC844879	16Sr XXXVI-A	Naderali et al., 2017
16Sr XXXVII：Stylosanthes little leaf group	1	'Ca. Phytoplasma stylosanthis'	MT431550	16Sr XXXVII-A	Jardim et al., 2021
16Sr XXXVIII：Bogia coconut syndrome group	1	'Ca. Phytoplasma noviguineense'	LC228755	16Sr XXXVIII-A	Miyazaki et al., 2018
16Sr XXXIX：Palm lethal wilt group	1	'Ca. Phytoplasma dypsidis'	MT536195	16Sr XXXIX-A	Jones et al., 2021

注：'Ca. Phytoplasma australasia'最初由White等（1998）描述。后来因为它的16S rRNA基因序列与'Ca. Phytoplasma aurantifolia'相似性为99.5%，且没有证据表明它代表了一个生态分离的种群（IRPCM, 2004），被从IRPCM列出的植原体候选'Ca. Phytoplasma'中移除。'Ca. Phytoplasma australasia'被错误地纳入2022年指南（Bertaccini et al., 2022），应予以删除。

表2-2 植原体基因组全图或草图信息表（Wei & Zhao, 2022）

植原体名称	株系	16Sr组分类	寄主症状	采样地	参考文献或GenBank上传信息	拼接序列号	拼接全长/bp	拼接水平	上传日期
'Catharanthus roseus' aster yellows phytoplasma	De Villa	I-B	maize bushy stunt-like	South Africa	Coetzee et al. deposited	GCF_004214875.1	603 949	Complete Genome	2019-2-20
'Chrysanthemum coronarium' phytoplasma	OY-V	I-B	onion yellows	Japan	Kakizawa et al., 2014	GCF_000744065.1	739 592	Contig	2014-8-14
'Cynodon dactylon' phytoplasma	LW01	XIV-A	bermuda grass white leaf	India	Kirdat et al., 2020	GCF_009268075.1	483 935	Scaffold	2019-10-22
'Echinacea purpurea' witches'-broom phytoplasma	NCHU2014	II-A	purple coneflower witches'-broom	China (Taiwan)	Chang et al., 2015	GCF_001307505.1	545 427	Contig	2015-10-7
'Fragaria × ananassa' phyllody phytoplasma	StrPh-Cl	XIII-F	strawberry phyllody	Chile	Cui et al., 2019	GCF_018274325.1	627 584	Contig	2021-5-4
'Parthenium hysterophorus' phyllody phytoplasma	PR34	II-new subgroup	santa-maria phyllody	India	Kirdat deposited	GCF_015100165.1	740 170	Contig	2020-10-29
'Parthenium sp.' Phyllody phytoplasma	PR08	II-D	santa-maria phyllody	India	Kirdat deposited	GCF_015239935.1	586 816	Contig	2021-5-10
'Santalum album' aster yellows phytoplasma	SW86	I-B	sandalwood spike	India	Tiwarekar deposited	GCF_018283495.1	554 025	Contig	2021-5-5
Aster yellows witches'-broom phytoplasma AYWB	AYWB	I-A	aster yellows witches'-broom in lettuce	USA	Bai et al., 2006	GCF_000012225.1	723 970	Complete Genome	2006-11-1

（续表）

植原体名称	株系	16Sr组分类	寄主症状	采样地	参考文献或GenBank上传信息	拼接序列号	拼接全长/bp	拼接水平	上传日期
Ca. Phytoplasma aurantifolia	WBDL	II-C	lime witches'-broom phytoplasma in periwinkle	Oman	Foissac and Carle deposited	GCF_002009625.1	474 669	Contig	2017-3-2
Ca. Phytoplasma luffae	NCHU2019	VIII-A	loofah witches'-broom	China (Taiwan)	Huang et al., 2022	GCF_018024475.1	769 143	Complete Genome	2021-4-16
Ca. Phytoplasma mali	AT	X-A	apple proliferation	NA	Kube et al., 2008	GCF_000026205.1	601 943	Complete Genome	2008-7-4
Ca. Phytoplasma	ChTYXIII-Mo	XIII-G	chinaberry yellowing	Argentina	Fernández et al., 2021	GCF_016876135.2	751 949	Contig	2021-4-14
Ca. Phytoplasma oryzae	NGS-S10	XI-A	napier grass stunt	Kenya	Fischer et al., 2016	GCF_003263355.1	484 488	Contig	2018-6-25
Ca. Phytoplasma phoenicium	SA213	XI-D	almond witches'-broom	Lebanon	Quaglino et al., 2015	GCF_001189415.1	345 965	Contig	2015-7-30
Ca. Phytoplasma pini	MDPP	XXI-B	pine phytoplasma	USA	Cai et al., 2020	GCF_007821455.1	474 136	Contig	2019-8-1
Ca. Phytoplasma pruni	ChTDIII	III-B	China-tree decline	Argentina	Fernández et al., 2020	GCF_013391955.1	790 517	Contig	2020-7-8
Ca. Phytoplasma pruni	CX	III-A	stone fruit tree decline	NA	Lee et al., 2015	GCF_001277135.1	598 511	Contig	2015-9-1
Ca. Phytoplasma sacchari	SCGS	XI-B	sugarcane grassy shoot	India	Kirdat et al., 2020	GCF_009268105.1	505 173	Contig	2019-11-4
Ca. Phytoplasma solani	SA-1	XII-A	bois noir in periwinkle	NA	Music et al., 2019	GCF_003698095.1	821 322	Contig	2018-10-30

(续表)

植原体名称	株系	16Sr组分类	寄主症状	采样地	参考文献或 GenBank 上传信息	拼接序列号	拼接全长/bp	拼接水平	上传日期
Ca. Phytoplasma solani	284/09	XII-A	stolbur phytoplasma (in tobacco and parsley)	NA	Mitrović et al., 2014	GCF_000970375.1	570 238	Chromosome	2013-10-20
Ca. Phytoplasma sp. AldY-WA1	AldY-WA1	V-A	alder yellows	USA	Cai et al., 2022	GCF_020312115.1	457 625	Scaffold	2021-10-6
Ca. Phytoplasma tritici	WBD	I-C	00420042pe blue dwarf	China	Chen et al., 2014	GCF_000495255.1	611 462	Contig	2013-11-1
Ca. Phytoplasma ziziphi	Jwb-nky	V-B	jujube witches'-broom	China	Wang et al., 2018	GCF_003640545.1	750 803	Complete Genome	2018-10-12
Chrysanthemum yellows phytoplasma	CYP	I-B	chrysanthemum yellows	Italy	Pacifico et al., 2015	GCF_000803325.1	659 699	Contig	2014-12-18
Hydrangea phyllody phytoplasma	HP	I-D	hydrangea phyllody	Japan	Nijo et al., 2021	GCF_018327665.1	597 775	Contig	2021-4-28
Italian clover phyllody phytoplasma str. MA1	MA1	III-B	Italian clover phyllody (in periwinkle)	Italy	Saccardo et al., 2012	GCF_000300695.1	597 245	Contig	2012-10-1
Maize bushy stunt phytoplasma	M3	I-B	maize bushy stunt	Brazil	Orlovskis et al., 2017	GCF_001712875.1	576 118	Complete Genome	2016-8-25
Milkweed yellows phytoplasma str. MW1	MW1	III-F	milkweed yellows (in periwinkle)	Italy	Saccardo et al., 2012	GCF_000309485.1	583 806	Contig	2012-10-1
Mulberry dwarf phytoplasma	MDGZ-01	I-B	mulberry dwarf	China	Luo et al., 2022	GCF_020714625.1	622 358	Complete Genome	2021-11-2

(续表)

植原体名称	株系	16Sr组分类	寄主症状	采样地	参考文献或GenBank上传信息	拼接序列号	拼接全长/bp	拼接水平	上传日期
New Jersey aster yellows phytoplasma	NJAY	I -A	new Jersey aster yellows（in periwinkle）	USA	Sparks et al., 2018	GCA_002554195.1	652 092	Contig	2017-10-16
Periwinkle leaf yellowing phytoplasma	DY2014	I -B	periwinkle leaf yellowing	China (Taiwan)	Cho et al., 2019	GCA_005093185.1	824 596	Contig	2019-5-2
'Brassica napus' phytoplasma	TW1	I -new subgroup	rapeseed stunting and virescence	Canada	Town et al., 2018	GCA_003181115.1	743 598	Contig	2018-5-31
'Elaeagnus angustifolia' witches'-broom phytoplasma	TBZ1	I -new subgroup	russian olive tree witches'-broom	Iran	Azizpour et al. deposited	GCA_018598675.1	833 199	Contig	2021-5-30
Onion yellows phytoplasma	OY	I -B	onion yellows（in chrysanthemum）	Japan	Oshima et al., 2004	GCA_000009845.1	853 092	Complete Genome	2003-12-9
Paulownia witches'-broom phytoplasma	Zhengzhou	I -D	paulownia witches'-broom	China	Cao et al., 2021	GCF_019396865.1	891 641	Complete Genome	2021-7-29
Peanut witches'-broom phytoplasma NTU2011	NTU2011	II -A	peanut witches'-broom（in periwinkle）	China (Taiwan)	Chung et al., 2013	GCF_000364425.1	566 694	Contig	2013-3-26
Poinsettia branch-inducing phytoplasma str. JR1	JR1	III -H	poinsettia branch-inducing（in periwinkle）	Italy	Saccardo et al., 2012	GCF_000309465.1	631 440	Contig	2012-10-1
Rice orange leaf phytoplasma	LD1	IX-A	rice orange leaf	China	Zhu et al., 2017	GCF_001866375.1	599 264	Contig	2016-11-4

（续表）

植原体名称	株系	16Sr组分类	寄主症状	采样地	参考文献或GenBank上传信息	拼接序列号	拼接全长/bp	拼接水平	上传日期
Sesame phyllody phytoplasma	SS02	II-A or II-D	sesame phyllody	India	Ranebennur et al., 2022	GCF_018390775.1	536 153	Contig	2021-5-17
Ca. Phytoplasma australiense		XII-B	maintained in periwinkle	Australia	Tran-Nguyen et al., 2008	GCA_000069925.1	879 959	Complete Genome	2008-4-2
Strawberry lethal yellows phytoplasma（CPA）	NZSb11	XII-B variant	strawberry lethal yellows	Australia and New Zealand	Andersen et al., 2013	GCF_000397185.1	959 779	Complete Genome	2013-5-16
Texas Phoenix palm phytoplasma	Flo-TPPD	IV-D	texas Phoenix Palm decline	USA	Bao et al. deposited	GCF_005774685.1	744 506	Contig	2019-5-23
Vaccinium witches'-broom phytoplasma str. VAC	VAC	III-F	vaccinium witches'-broom（in periwinkle）	Italy	Saccardo et al., 2012	GCF_000309405.1	647 754	Contig	2012-10-1
Ca. Phytoplasma sp.	Tabriz.2	I-B	Elaeagnus sp.（symptoms not described）	Iran	Zirak et al. deposited	GCA_019841745.1	762 261	Contig	2021-8-24
Ca. Phytoplasma trifolii-related	CBPPT1	VI-A	potato purple top（in periwinkle）	USA	Wei et al. deposited	PRJNA839414	514 536	Contig	2022-5-18
Florescence dorée (FD) phytoplasma	CH	V-A	florescence dorée（in insect vector Scaphoideus titanus）	Switzerland	Debonneville et al., 2022	PRJNA838420	654 223	Complete Genome	2022-6-27

参考文献

蔡红，陈惠，李凡，等，2003. 长春花黄化植原体核糖体蛋白基因片段序列分析[J]. 微生物学通报，30（1）：34-37.

AL-SAADY N A, KHAN A J, CALARI A, et al.，2008.'Candidatus Phytoplasma omanense', associated with witches'-broom of Cassia italica（Mill.）Spreng. in Oman[J]. International Journal of Systematic and Evolutionary Microbiology，58：461-466.

ANDERSEN M T, LIEFTING L W, HAVUKKALA I, et al.，2013. Comparison of the complete genome sequence of two closely related isolates of 'Candidatus Phytoplasma australiense' reveals genome plasticity[J]. BMC Genomics，14：529.

AROCHA Y, ANTESANA O, MONTELLANO E, et al.，2007.'Candidatus Phytoplasma lycopersici', a phytoplasma associated with 'hoja de perejil' disease in Bolivia[J]. International Journal of Systematic and Evolutionary Microbiology，57：1704-1710.

AROCHA Y, LOPEZ M, PINOL B, et al.，2005.'Candidatus Phytoplasma graminis' and 'Candidatus Phytoplasma caricae', two novel phytoplasmas associated with diseases of sugarcane, weeds and papaya in Cuba[J]. International Journal of Systematic and Evolutionary Microbiology，55：2451-2463.

BAI X, ZHANG J, EWING A, et al.，2006. Living with genome instability：The adaptation of phytoplasmas to diverse environments of their insect and plant hosts[J]. Journal of Bacteriology，188：3682-3696.

BERTACCINI A, AROCHA-ROSETE Y, CONTALDO N, et al.，2022. Revision of the 'Candidatus Phytoplasma' species description guidelines[J]. International Journal of Systematic and Evolutionary Microbiology，72：005353.

CAI W, NUNZIATA S O, SRIVASTAVA S K, et al.，2022. Draft genome sequence resource of AldY-WA1, a phytoplasma strain associated with alder yellows of Alnus rubra in Washington, U. S. A[J]. Plant Disease，106：1971-1973.

CAI W, SHAO J, ZHAO Y, et al.，2020. Draft genome sequence of 'Candidatus Phytoplasma pini'-related strain MDPP：A resource for comparative genomics of gymnosperm-infecting phytoplasmas[J]. Plant Disease，104：1009-1010.

CAO Y, SUN G, ZHAI X, et al.，2021. Genomic insights into the fast growth of paulownias and the formation of Paulownia witches'-broom[J]. Molecular Plant，14：1668-1682.

CHANG S H, CHO S T, CHEN C L, et al.，2015. Draft genome sequence of a 16SrⅡ-A subgroup phytoplasma associated with purple coneflower（Echinacea purpurea）witches'-broom disease in Taiwan[J]. Genome Announcements，3：e01398-15.

CHEN W, LI Y, WANG Q, et al., 2014. Comparative genome analysis of wheat blue dwarf phytoplasma, an obligate pathogen that causes wheat blue dwarf disease in China[J]. PLoS ONE, 9: e96436.

CHO S T, LIN C P, KUO C H, 2019. Genomic characterization of the periwinkle leaf yellowing (PLY) phytoplasmas in Taiwan[J]. Frontiers in Microbiology, 10: 2194.

CHUNG W C, CHEN L L, LO W S, et al., 2013. Comparative analysis of the peanut witches'-broom phytoplasma genome reveals horizontal transfer of potential mobile units and effectors[J]. PLoS ONE, 8: e62770.

CUI W, QUIROGA N, CURKOVIC S T, et al., 2019. Detection and identification of 16SrXIII-F and a novel 16SrXIII phytoplasma subgroups associated with strawberry phyllody in Chile[J]. European Journal of Plant Pathology, 155: 1039-1046.

DAVIS R E, DALLY E L, GUNDERSEN D E, et al., 1997. "Candidatus Phytoplasma australiense" a new phytoplasma taxon associated with Australian grapevine yellows[J]. International Journal of Systematic Bacteriology, 47: 262-269.

DAVIS R E, HARRISON N A, ZHAO Y, et al., 2016. 'Candidatus Phytoplasma hispanicum', a novel taxon associated with Mexican periwinkle virescence disease of Catharanthus roseus[J]. International Journal of Systematic and Evolutionary Microbiology, 66: 3463-3467.

DAVIS R E, ZHAO Y, WEI W, et al., 1999. 'Candidatus Phytoplasma luffae', a novel taxon associated with witches'-broom disease of loofah, Luffa aegyptica Mill[J]. International Journal of Systematic and Evolutionary Microbiology, 67: 3127-3133.

DAVIS R E, ZHAO Y, DALLY E L, et al., 2012. 'Candidatus Phytoplasma sudamericanum', a novel taxon, and strain PassWB-Br4, a new subgroup 16SrIII-V phytoplasma, from diseased passion fruit (Passiflora edulis f. flavicarpa Deg.) [J]. International Journal of Systematic and Evolutionary Microbiology, 62 Pt 4: 984-989.

DAVIS R E, ZHAO Y, DALLY E L, et al., 2013. 'Candidatus Phytoplasma pruni', a novel taxon associated with X-disease of stone fruits, Prunus spp.: Multilocus characterization based on 16S rRNA, secY, and ribosomal protein genes[J]. International Journal of Systematic and Evolutionary Microbiology, 63 Pt 2: 766-776.

DEBONNEVILLE C, MANDELLI L, BRODARD J, et al., 2022. The complete genome of the "Flavescence Dorée" phytoplasma reveals characteristics of low genome plasticity[J]. Biology, 11: 953.

DOI Y, TERANAKA M, YORA K, et al., 1967. Mycoplasma or PLT group like microorganisms found in the phloem elements of plants infected with mulberry dwarf, americanum', a phytoplasma associated with a potato purple top wilt disease complex[J]. International Journal of Systematic and Evolutionary Microbiology, 56: 1593-1597.

LEE I M, DAVIS R E, GUNDERSEN-RINDAL D E, 2000. Phytoplasma: phytopathogenic mollicutes[J]. Annual Review of Microbiology, 54 (1): 221-255.

LEE I M, GUNDERSEN-RINDAL D E, DAVIS R E, et al., 1998. Revised classification scheme of phytoplasmas based on RFLP analyses of 16S rRNA and ribosomal protein gene sequence[J]. International Journal of Systematic Bacteriology, 48 (7): 1153-1169.

LEE I M, MARTINI M, MARCONE C, et al., 2004a. Classification of phytoplasma strains in the elm yellows group (16SrⅤ) and proposal of 'Candidatus Phytoplasma ulmi' for the phytoplasma associated with elm yellows[J]. International Journal of Systematic and Evolutionary Microbiology, 54: 337-347.

LEE I M, GUNDERSEN-RINDAL D E, DAVIS R E, et al., 2004b. 'Candidatus Phytoplasma asteris', a novel phytoplasma taxon associated with aster yellows and related diseases[J]. International Journal of Systematic and Evolutionary Microbiology, 54: 1037-1048.

LEE I M, SHAO J, BOTTNER-PARKER K D, et al., 2015. Draft genome sequence of "Candidatus Phytoplasma pruni" strain CX, a plant-pathogenic bacterium[J]. Genome Announcement, 3: e01117-15.

LEE I M, BOTTNER-PARKER K D, ZHAO Y, et al., 2011. 'Candidatus Phytoplasma costaricanum', a novel phytoplasma associated with an emerging disease in soybean (Glycine max) [J]. International Journal of Systematic and Evolutionary Microbiology, 61: 2822-2826.

LI Y, PIAO C G, TIAN G Z, et al., 2014. Multilocus sequences confirm the close genetic relationship of four phytoplasmas of peanut witches'-broom group 16SrⅡ-A[J]. Journal of Basic Microbiology, 54 (8): 818-827.

LIEFTING L W, ANDERSEN M T, BEEVER R E, et al., 1996. Sequence heterogeneity in two 16S rRNA genes of Phormium yellow leaf phytoplasma[J]. Applied and Environmental Microbiology, 62 (9): 3133-3139.

LIM P O, SEARS B B, 1989. 16S rRNA sequence indicates that plant-pathogenic mycoplasma like organisms are evolutionarily distinct from animal mycoplasmas[J]. Journal of Bacteriology, 171 (11): 5901-5906.

LIM P O, SEARS B B, 1992. Evolutionary relationships of a plant-pathogenic mycoplasmalike organism and Acholeplasma laidlawii deduced from two ribosomal protein gene sequences[J]. Journal of Bacteriology, 174 (8): 2606-2611.

LUO L, ZHANG X, MENG F, et al., 2022. Draft genome sequences resources of mulberry dwarf phytoplasma strain MDGZ-01 associated with mulberry yellow dwarf（MYD）diseases[J]. Plant Disease, 106: 2239-2242.

MALEMBIC-MAHER S, SALAR P, FILIPPIN L, et al., 2011. Genetic diversity of European phytoplasmas of the 16SrⅤ taxonomic group and proposal of 'Candidatus Phytoplasma rubi' [J]. International Journal of Systematic and Evolutionary Microbiology, 61: 2129-2134.

MARCONE C, GIBB K S, STRETEN C, et al., 2004. 'Candidatus Phytoplasma spartii', 'Candidatus Phytoplasma rhamni' and 'Candidatus Phytoplasma allocasuarinae', respectively associated with spartium witches'-broom, buckthorn witches'-broom and allocasuarina yellows diseases[J]. International Journal of Systematic and Evolutionary Microbiology, 54: 1025-1029.

MARCONE C, SCHNEIDER B, SEEMÜLLER E, 2004. 'Candidatus Phytoplasma cynodontis', the phytoplasma associated with Bermuda grass white leaf disease[J]. International Journal of Systematic and Evolutionary Microbiology, 54: 1077-1082.

MARTINI M, LEE I M, BOTTNER K D, et al., 2007. Ribosomal protein gene-based phylogeny for finer differentiation and classification of phytoplasmas[J]. International Journal of Systematic and Evolutionary Microbiology, 57（9）: 2037-2051.

MARTINI M, MARCONE C, MITROVI'C J, et al., 2012. 'Candidatus Phytoplasma convolvuli', a new phytoplasma taxon associated with bindweed yellows in four European countries[J]. International Journal of Systematic and Evolutionary Microbiology, 62 Pt 12: 2910-2915.

MELAMED S, TANNE E, BEN-HAIM R, et al., 2003. Identification and characterization of phytoplasmal genes, employing a novel method of isolating phytoplasmal genomic DNA[J]. Journal of Bacteriology, 185（22）: 6513-6520.

MELLO A P O A, BEDENDO I P, CAMARGO L E A, 2006. Sequence heterogeneity in the 16S rDNA of tomato big bud phytoplasma belonging to group 16SrⅢ[J]. Phytopathology, 154（4）: 245-249.

MITROVI'C J, SIEWERT C, DUDUK B, et al., 2014. Generation and analysis of draft sequences of 'stolbur' phytoplasma from multiple displacement amplification templates[J]. Microbiology Physiology, 24: 1-11.

MURRAY R G E, SCHLEIFER K H, 1994. Taxonomic notes: A proposal for recording the properties of putative taxa of procaryotes[J]. International Journal of Systematic and Evolutionary Microbiology, 44: 174-176.

MUSIC M S, SAMARZIJA I, HOGENHOUT S A, et al., 2019. The genome of 'Candidatus

Phytoplasma solani' strain SA-1 is highly dynamic and prone to adopting foreign sequences[J]. Systematic and Applied Microbiology, 42: 117-127.

MIYAZAKI A, SHIGAKI T, KOINUMA H, et al., 2018. 'Candidatus Phytoplasma noviguineense', a novel taxon associated with Bogia coconut syndrome and banana wilt disease on the island of New Guinea[J]. International Journal of Systematic and Evolutionary Microbiology, 68: 170-175.

MONTANO H G, DAVIS R E, DALLY E L, et al., 2001. 'Candidatus Phytoplasma brasiliense', a new phytoplasma taxon associated with hibiscus witches'-broom disease[J]. International Journal of Systematic and Evolutionary Microbiology, 51: 1109-1118.

NADERALI N, NEJAT N, VADAMALAI G, et al., 2017. 'Candidatus Phytoplasma wodyetiae', a new taxon associated with yellow decline disease of foxtail palm (Wodyetia bifurcata) in Malaysia[J]. International Journal of Systematic and Evolutionary Microbiology, 67: 3765-3772.

NEJAT N, VADAMALAI G, DAVIS R E, et al., 2013. 'Candidatus Phytoplasma malaysianum', a novel taxon associated with virescence and phyllody of Madagascar periwinkle (Catharanthus roseus)[J]. International Journal of Systematic and Evolutionary Microbiology, 63 Pt 2: 540-548.

NIJO T, IWABUCHI N, TOKUDA R, et al., 2021. Enrichment of phytoplasma genome DNA through a methyl-CpG binding domain-mediated method for efficient genome sequencing[J]. Journal of General Plant Pathology, 87: 154-163.

ORLOVSKIS Z, CANALE M C, HARYONO M, et al., 2017. A few sequence polymorphisms among isolates of Maize bushy stunt phytoplasma associate with organ proliferation symptoms of infected maize plants[J]. Annals of Botany, 119: 869-884.

OSHIMA K, KAKIZAWA S, NISHIGAWA H, et al., 2004. Reductive evolution suggested from the complete genome sequence of a plant-pathogenic phytoplasma[J]. Nature Genetics, 36: 27-29.

PACIFICO D, GALETTO L, RASHIDI M, et al., 2015. Decreasing global transcript levels over time suggest that phytoplasma cells enter stationary phase during plant and insect colonization[J]. Applied and Environment Microbiology, 81: 2591-2602.

QUAGLINO F, KUBE M, JAWHARI M, et al., 2015. 'Candidatus Phytoplasma phoenicium' associated with almond witches'-broom disease: From draft genome to genetic diversity among strain populations[J]. BMC Microbiology, 15: 148.

QUAGLINO F, ZHAO Y, CASATI P, et al., 2013. 'Candidatus Phytoplasma solani', a novel taxon associated with stolbur-and bois noir-related diseases of plants[J]. International

Journal of Systematic and Evolutionary Microbiology, 63: 2879-2894.

RANEBENNUR H, KIRDAT K, TIWAREKAR B, et al., 2022. Draft genome sequence of 'Candidatus Phytoplasma australasia', strain SS02 associated with sesame phyllody disease[J]. 3 Biotech, 12: 107.

SACCARDO F, MARTINI M, PALMANO S, et al., 2012. Genome drafts of four phytoplasma strains of the ribosomal group 16SrⅢ[J]. Microbiology, 158: 2805-2814.

ŠAFÁŘOVÁ D, ZEMANEK T, VALOVA P, et al., 2016. 'Candidatus Phytoplasma cirsii', a novel taxon from creeping thistle[Cirsium arvense (L.) Scop.][J]. International Journal of Systematic and Evolutionary Microbiology, 66: 1745-1753.

SAWAYANAGI T, HORIKOSHI N, KANEHIRA T, et al., 1999. 'Candidatus Phytoplasma japonicum', a new phytoplasma taxon associated with Japanese Hydrangea phyllody[J]. International Journal of Systematic and Evolutionary Microbiology, 49: 1275-1285.

SCHNEIDER B, GIBB K S, SEEMÜLLER E, 1997. Sequence and RFLP analysis of the elongation factor Tu gene used in differentiation and classification of phytoplasmas[J]. Microbiology, 143 (10): 3381-3389.

SCHNEIDER B, TORRES E, MARTÍN M P, et al., 2005. 'Candidatus Phytoplasma pini', a novel taxon from Pinus silvestris and Pinus halepensis[J]. International Journal of Systematic and Evolutionary Microbiology, 55: 303-307.

SEEMÜLLER E, MARCONE C, LAUER U, et al., 1998. Current status of molecular classification of the phytoplasma[J]. Journal of Plant Pathology, 80 (1): 3-26.

SEEMÜLLER E, SCHNEIDER B, 2004. 'Candidatus Phytoplasma mali', 'Candidatus Phytoplasma pyri' and 'Candidatus Phytoplasma prunorum', the causal agents of apple proliferation, pear decline and European stone fruit yellows, respectively[J]. International Journal of Systematic and Evolutionary Microbiology, 54: 1217-1226.

SPARKS M E, BOTTNER-PARKER K D, GUNDERSEN-RINDAL D E, et al., 2018. Draft genome sequence of the New Jersey aster yellows strain of 'Candidatus Phytoplasma asteris'[J]. PLoS One, 13: e0192379.

TOWN J R, WIST T, PEREZ-LOPEZ E, et al., 2018. Genome sequence of a plant-pathogenic bacterium, "Candidatus Phytoplasma asteris" strain TW1[J]. Microbiology Resource Announcement, 7: e01109-18.

TRAN-NGUYEN L T, KUBE M, SCHNEIDER B, et al., 2008. Comparative genome analysis of "Candidatus Phytoplasma australiense" (subgroup tuf-Australia I; rp-A) and "Ca. Phytoplasma asteris" strains OY-M and AY-WB[J]. Journal of Bacteriology, 190: 3979-3991.

VALIUNAS D, JOMANTIENE R, DAVIS R E, 2013. Evaluation of the DNA-dependent

RNA polymerase β-subunit gene (*rpoB*) for phytoplasma classification and phylogeny[J]. International Journal of Systematic and Evolutionary Microbiology, 63(10): 3904-3914.

VALIUNAS D, STANIULIS J, DAVIS R E, 1999. 'Candidatus Phytoplasma fragariae', a novel phytoplasma taxon discovered in yellows diseased strawberry, Fragaria × ananassa[J]. International Journal of Systematic and Evolutionary Microbiology, 56: 277-281.

VERDIN E, SALAR P, DANET J L, et al., 2003. 'Candidatus Phytoplasma phoenicium' sp. nov., a novel phytoplasma associated with an emerging lethal disease of almond trees in Lebanon and Iran[J]. International Journal of Systematic and Evolutionary Microbiology, 53: 833-838.

WANG J, SONG L, JIAO Q, et al., 2018. Comparative genome analysis of jujube witches'-broom Phytoplasma, an obligate pathogen that causes jujube witches'-broom disease[J]. BMC Genomics, 19: 689.

WEI W, DAVIS R E, LEE I M, et al., 2007. Computer-simulated RFLP analysis of 16S rRNA genes: identification of ten new phytoplasma groups[J]. International Journal of Systematic and Evolutionary Microbiology, 57(8): 1855-1867.

WEI W, ZHAO Y, 2022. Phytoplasma taxonomy: nomenclature, classification, and identification[J]. Biology, 11: 1119.

WHITE D T, BLACKALL L L, SCOTT P T, et al., 1998. Phylogenetic positions of phytoplasmas associated with dieback, yellow crinkle and mosaic diseases of papaya, and their proposed inclusion in 'Candidatus Phytoplasma australiense' and a new taxon, 'Candidatus Phytoplasma australasia'[J]. International Journal of Systematic and Evolutionary Microbiology, 48: 941-951.

WIN N K K, LEE S Y, BERTACCINI A, et al., 2013. 'Candidatus Phytoplasma balanitae' associated with witches'-broom disease of Balanites triflora[J]. International Journal of Systematic and Evolutionary Microbiology, 63(Pt 2): 636-640.

ZHAO Y, SUN Q, WEI W, et al., 2009. 'Candidatus Phytoplasma tamaricis', a novel taxon discovered in witches'-broom-diseased salt cedar (*Tamarix chinensis* Lour.)[J]. International Journal of Systematic and Evolutionary Microbiology, 59: 2496-2504.

ZHU Y, HE Y, ZHENG Z, et al., 2017. Draft genome sequence of rice orange leaf phytoplasma from Guangdong, China[J]. Genome Announcement, 5: e00430-17.

ZREIK L, CARLE P, BOVE J M, et al., 1995. Characterization of the mycoplasmalike organism associated with witches'-broom disease of lime and proposition of a "Candidatus" taxon for the organism, "Candidatus Phytoplasma aurantifolia"[J]. International Journal of Systematic Bacteriology, 45: 449-453.

第三章

植原体及其病害检测诊断技术

植原体尚难分离培养，无法通过科赫法则（Koch's postulates）对感染植原体的植物病害进行病原验证，植原体病害发现初期，根据植物的病害症状进行初步的判断，如丛枝、黄化、小叶、花器变态等症状，再进一步通过电子显微镜进行切片观察，采取注射四环素类的抗生素确定是不是由于植原体引起的病害。随着分子生物学的发展，借助分子生物学方法对植原体进行检测的技术相继被开发出现，如直接PCR、巢式PCR、LAMP、数字PCR等，这些检测方法的灵敏度以及诊断的准确性也越来越高，成为了普遍被接受而且应用最广泛的植原体分子诊断方法。

一、电子显微镜检测技术

电子显微镜技术，可直接对植原体形态和结构进行观察，是确定植原体侵染最直接的证据。国内外学者也借助该技术，证实了多种植物病害体内含有大量的植原体（金开璇等，1982；王纯利等，1992）。1967年，日本学者Doi等通过电子显微镜对黄化型桑树萎缩病等病害的超薄切片进行观察，发现了植原体的存在，而这一技术作为植原体检测最经典的方法，在植原体的发现以及病害的诊断过程中发挥了不可替代的作用。但是由于植原体病原菌在植物体内分布不均匀，寄主植物在不同季节、不同组织部位的植原体含量有很大的差别，特别是一些含量低的寄主体内，植原体病原菌的捕捉率很低，而且电镜技术本身对设备有较高的要求，如检测过程费时、耗材昂贵、操作技术要求高等不足，使电子显微镜技术在现代植原体诊断和检测中使用得越来越少（王圣洁，2017）。

二、组织化学技术

植原体的入侵常常导致寄主植物体内产生一些特异性物质代谢变化，利用荧光染料对这些特异性代谢物质进行染色，并通过光学显微镜观察，基于这一代谢特点，研究人员开增产物为模板，极大提高了扩增效率及检测灵敏度（王圣洁，2017；车海彦等，2010；于少帅等，2021）。

根据Gundersen等（1996）和Lee等（1993）报道的引物对扩增植原体16S rDNA基因片段序列。R16mF2/R16mR1作为第一引物对，R16F2n/R16R2作为第二引物对进行巢式PCR扩增。引物对序列如下。

  R16mF2：5′-CATGCAAGTCGAACGGA-3′；

  R16mR1：5′-CTTAACCCCAATCATCGAC-3′；

  R16F2n：5′-ACGACTGCTAAGACTGG-3′；

  R16R2：5′-GCGGTGTGTACAAACCCCG-3′。

PCR反应模板：直接PCR以提取的总DNA为模板，巢式PCR以直接PCR产物稀释50倍后为模板。PCR反应均以健康植物材料总DNA为阴性对照和无菌双蒸水为空白对照。反应体系为25 μL，含有10×PCR buffer 2.5 μL、10 μmol/L引物各1 μL、10 mmol/L dNTP 2 μL、0.125 μL *Taq* DNA聚合酶5 U/μL（大连TaKaRa公司）、1 μL模板DNA、无菌水18.375 μL。PCR反应条件为94℃，预变性2 min；94℃变性1 min；45℃退火45 s，72℃延伸1 min；变性、退火、延伸共30个循环，最后72℃延伸10 min。PCR产物在1%琼脂糖凝胶中电泳，经溴化乙锭染色后，使用UVI凝胶成像系统观察。PCR扩增出的目标产物由TaKaRa Agarose Gel DNA Purification Kit Ver.2.0纯化回收，将目标产物与pMD18-T Simple载体连接，转化到大肠杆菌DH5α中，利用PCR反应筛选阳性克隆。

五、实时荧光定量PCR技术

实时荧光定量PCR是一种通过荧光物质对PCR扩增的反应过程进行监测，并通过内参或外参实现对目标DNA片段定量的方法。Christensen等（2004）基于荧光定量PCR技术对感染植原体的植物组织进行定量检测，发现植原体主要集中在叶部，在叶柄和茎部植原体含量较少，证实了植原体在组织中分布的不均性。Torres等（2005）基于16S rRNA基因建立了16SrX组的不同植原体株系的实时荧光定量PCR扩增体系，可以同时检测欧洲多种果树植原体病害。实时荧光定量PCR技术于1996年由美国Applied Biosystems公司推出，该技术是利用荧光信号伴随着PCR产物的增加而增强的原理，在PCR扩增过程中，连续监测反应体系中荧光信号的变化。荧光定量PCR所使用的荧光化学可分为荧光探针（*TaqMan*、*TaqMan* MGB和双探针）和荧光染料（SYBR Green I）2种。车海彦等（2009）基于*TaqMan* MGB探针建立了槟榔黄化植原体的实时荧光PCR检测体系。任争光等（2015）利用SYBR® Green I染料建立了枣疯病植原体实时荧光定量PCR检测体系。

车海彦等（2009）以植原体16SrDNA为目标基因进行引物、探针设计，从NCBI中下载所有已知植原体的16S rRNA基因序列，用ClustalX1.81软件进行比对，获得16SrI组植原体组内序列保守而与其他组间序列有稳定性突变的区域，然后使用Primer version 3.0软件设计实时荧光PCR引物/TaqMan MGB探针，利用NCBI中的Blast程序（http://blast.ncbi.nlm.nih.gov/Blast）分别将正向和反向引物进行比对，未发现除16SrⅠ组之外的可以与此对引物配对并引起扩增的物种。序列如下所示。

引物序列：

上游引物，PbLF：5′-GCGAAGGCGGCTTGCT-3′（序列表中序列2）；

下游引物，PbLR：5′-TTTGCTCCCCACGCTTTC-3′（序列表中序列3）。

探针序列：

bFprobe：5′-FAM-CTTTACTGACGCTGAGGCA-MGBNFQ-3′（FAM指荧光报告基团，MGB指小沟结合物，NFQ指非荧光淬灭基团）。

实时荧光PCR反应体系为：*TaqMan* Universal PCR Master Mix 12.5 μL；引物PbLF 650 nmol/L；引物PbLR 650 nmol/L；探针bFprobe 230 nmol/L；模板DNA 1 μL；补无核酸酶水（Nuclease-free water）至25 μL。

六、环介导等温基因扩增技术

环介导等温扩增技术（loop-mediated isothermal amplification，LAMP）是Notomi等（2000）开发的一种恒温核酸扩增方法，是一种新型的分子诊断方法。该方法具有特异性强、操作简单、不需要复杂仪器、检测周期短和检测结果可视化的特点，非常适合作为基层和现场的病害诊断方法进行推广和使用。Sugawara等（2012）基于*groEL*基因开发的LAMP引物可以检测翠菊黄化等植原体株系。Obura等（2011）基于16S rRNA基因设计的LAMP引物能够检测不同组的不同植原体株系，而多数引物仅能检测相同组内的植原体。王圣洁等（2017b）基于*tuf*基因序列开发的LAMP引物适合于泡桐丛枝、苦楝丛枝等植原体株系的检测。Nair等（2016）基于印度16SrⅪ组槟榔黄化植原体16S rRNA基因建立适合检测印度16SrⅪ组槟榔黄化植原体的LAMP检测体系。针对热区棕榈作物，于少帅等（2020，2023）建立了针对槟榔、椰子等植原体病害的环介导等温基因扩增检测技术。

Yu等（2020）建立了针对我国16SrⅠ组槟榔黄化植原体的快速高效检测体系，验证检测体系的特异性、灵敏度和稳定性，并用针对我国16SrⅠ组槟榔黄化植原体LAMP技术对我国海南保亭、屯昌、万宁等槟榔黄化病样品进行LAMP检测。用环介导等温扩增法引物在线设计软件PrimerExplorer V4（http://primerexplorer.jp/e/）设计引物，按照LAMP引物设计原则，针对16SrⅠ组槟榔黄化植原体16S rRNA基因的保守序列，设计多套LAMP引物，从中筛选出特异性好的引物组。LAMP扩增的引物包括正向外引物F3、反向外引物B3、正向内引物FIP（FIC+F2）、反向内引物BIP（BIC+B2）。引物由生工生物工程（上海）股份有限公司按照HPLC纯度级别合成，引物序列如表3-2所示。参考荣研生物科技（中国）有限公司所提供的恒温扩增试剂盒，建立槟榔黄化植原体LAMP扩增体系。槟榔黄化植原体扩增体系25.0 μL，其中2×反应缓冲液（RM）12.5 μL，*Bst* DNA聚合酶1.0 μL，引物（40 pmol/μL FIP/BIP、20 pmol/μL LF/LB、5 pmol/μL F3/B3）各1 μL，荧光目视检测试剂（FD）1 μL，自制备DNA模板2 μL，去离子水补齐至25 μL。64℃恒温反应60 min。LAMP阳性扩增结果可以通过2种方法鉴别：一是将扩增反应管置于恒温荧光检测仪或荧光PCR仪中，实时读取荧光信号，如果出现"S"形扩增曲线则判定为阳性，即待测样品中含有植原体，如果无"S"形扩增曲线则判定为阴性，即待测样品中无植原体；二是反应前加入荧光目视检测试剂FD显色剂（钙黄绿素-氯化锰溶液），则可根据颜色变化判断结果，如果扩增产物变为翠绿色则判定为阳性，即待测样品中含有植原体，如果扩增产物仍为橘黄色则判定为阴性，即待测样品中无植原体（Yu et al.，2020）。

表3-2　LAMP与PCR扩增引起序列信息（Yu et al.，2020）

引物名称	引物序列（5′-3′）
16SrDNA-F3-2	GCATGGTTGTCGTCAGCT
16SrDNA-B3-2	GCCAAAAACTTGCGCTTCA
16SrDNA-FIP-2	GGCAGTCTTGCTAAAGTCCCCACGATGTTGGGTTAAGTCCCGC
16SrDNA-BIP-2	ACGACGTCAAATCATCATGCCCCGCTACCCTTTGTAACAGCCAT
16SrDNA-LF-2	AATAAGGGTTGCGCTCGTT
16SrDNA-LB-2	ATGACCTGGGCTACAAACGT
16SrDNA-F3-3	GCTGAAGCGCAAGTTTTGG
16SrDNA-B3-3	ACCTTAGACGGTTCCCTCTTC
16SrDNA-FIP-3	GCGACATGCTGATTCGCGATTACTCAGTTCGGATTGAAGTCTGC
16SrDNA-BIP-3	ACGTTCTCGGGGTTTGTACACACTTGCGAAGTTAGGCCACC
16SrDNA-LF-3	TCCAACTTCATGAAGTCGAGTT
16SrDNA-LB-3	CGCCCGTCAAACCACGA
P1	AAGAGTTTGATCCTGGCTCAGGATT
P7	CGTCCTTCATCGGCTCTT
R16mF2	CATGCAAGTCGAACGGA
R16mR1	CTTAACCCCAATCATCGAC

Yu等（2020）以万宁槟榔黄化植原体样品和健康样品为阳性（WN+）和阴性（WN-）实验材料，以去离子水（CK）为对照，筛选以16S rRNA基因作为靶标设计的多

套LAMP引物组，参考恒温扩增试剂盒所提供的反应程序，建立槟榔黄化植原体LAMP检测体系，确定LAMP检测的特异性。由图3-1、图3-2、图3-3可知，基于16S rRNA基因设计的2套槟榔黄化植原体LAMP引物组：16SrDNA-2 LAMP引物组和16SrDNA-3 LAMP引物组，通过恒温扩增技术扩增70 min，2套LAMP引物组的槟榔黄化植原体样品和LAMP试剂盒自带的阳性对照（P+）均出现"S"形曲线，反应前加入显色液后反应结果均变成翠绿色，而槟榔健康样品、LAMP试剂盒自带的阴性对照（N-）和去离子水对照均未出现"S"形扩增曲线和变色反应。槟榔黄化植原体LAMP检测体系从槟榔黄化植原体基因组DNA加入至阳性结果出现可以在40 min内完成，由此实现了对槟榔黄化植原体染病组织的快速检测。

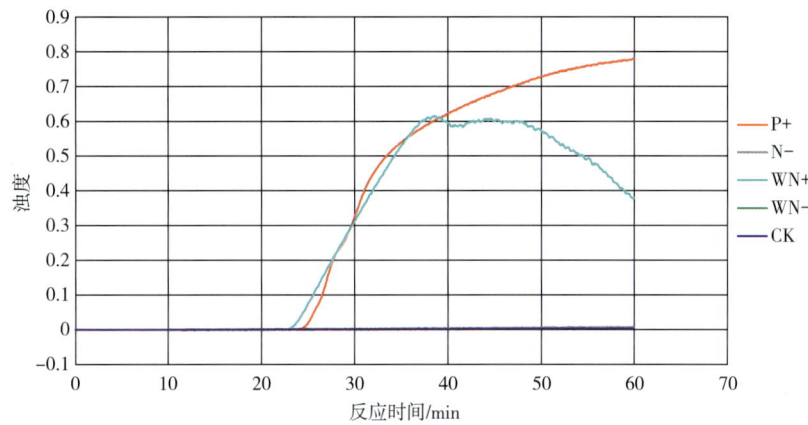

P+—LAMP试剂盒自带的阳性对照；N-—LAMP试剂盒自带的阴性对照；
WN+—槟榔黄化植原体样品；WN-—槟榔健康样品；CK—去离子水。

图3-1　槟榔黄化植原体16SrDNA-2 LAMP引物组实时扩增曲线图（Yu et al.，2020）

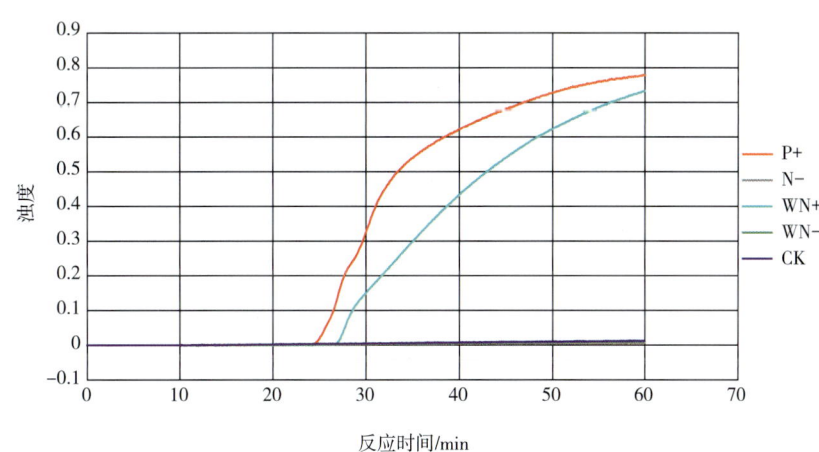

P+—LAMP试剂盒自带的阳性对照；N-—LAMP试剂盒自带的阴性对照；
WN+—槟榔黄化植原体样品；WN-—槟榔健康样品；CK—去离子水。

图3-2　槟榔黄化植原体16SrDNA-3 LAMP引物组实时扩增曲线图（Yu et al.，2020）

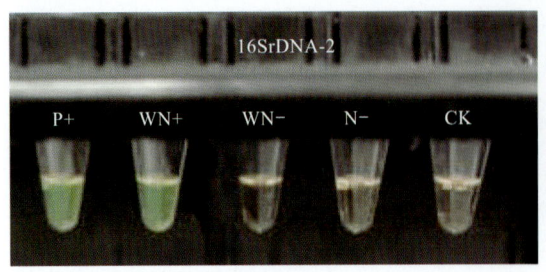 

P+——LAMP试剂盒自带的阳性对照；N-——LAMP试剂盒自带的阴性对照；
WN+——槟榔黄化植原体样品；WN-——槟榔健康样品；CK—去离子水。

图3-3 槟榔黄化植原体16SrDNA-2和16SrDNA-3 LAMP引物组变色反应（Yu et al.，2020）

将基于pUC57质粒载体构建的16SrDNA重组质粒溶解为浓度20 ng/μL的原液，以10倍梯度进行稀释，稀释后的稀释液作为LAMP检测的模板，通过实时扩增曲线和变色反应进行判读。于少帅等（2020）研究结果表明，以200 ag/μL稀释液为模板，16SrDNA-2和16SrDNA-3两套LAMP引物组均有"S"形实时扩增曲线和变色反应，而稀释10倍后以20 ag/μL为模板，16SrDNA-2和16SrDNA-3 2套LAMP引物组均无"S"形实时扩增曲线和变色反应。由此可见槟榔黄化植原体16SrDNA-2和16SrDNA-3 两套LAMP引物组特异性良好，灵敏度最低检出限均可达到200 ag/μL，约53个拷贝数，如图3-4、图3-5、图3-6所示。

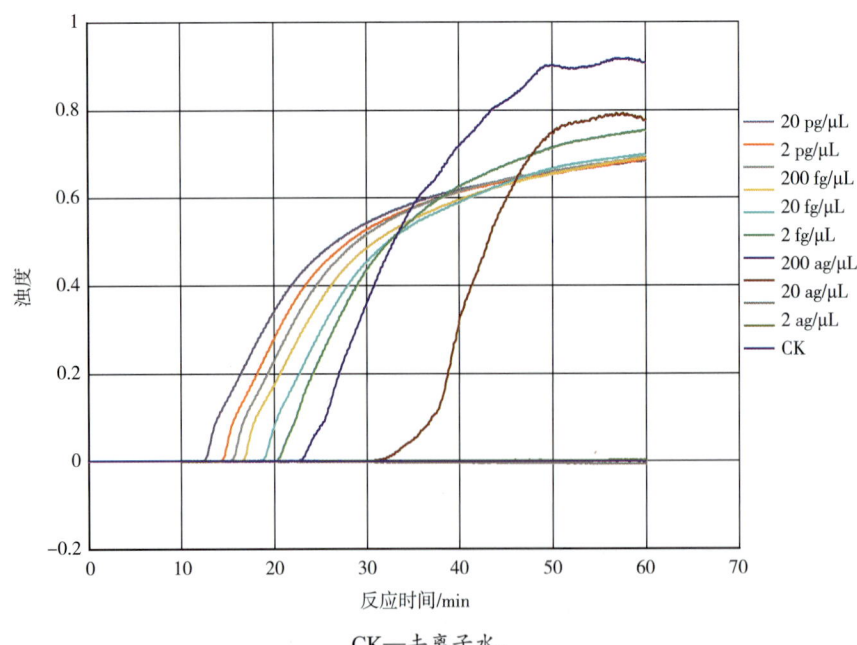

CK—去离子水。

图3-4 槟榔黄化植原体16SrDNA-2 LAMP引物组灵敏度实时扩增曲线图（Yu et al.，2020）

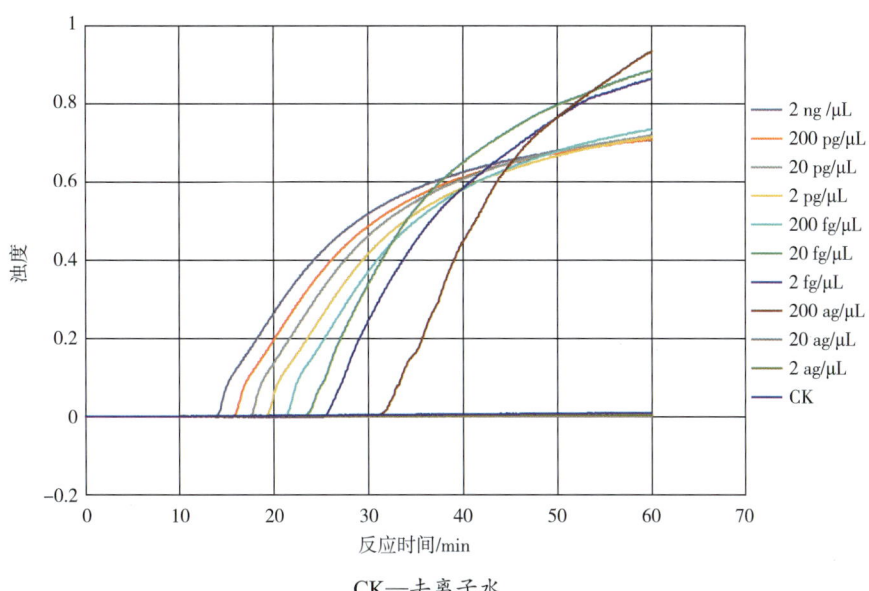

CK—去离子水。

图3-5　槟榔黄化植原体16SrDNA-3 LAMP引物组灵敏度实时扩增曲线图（Yu et al., 2020）

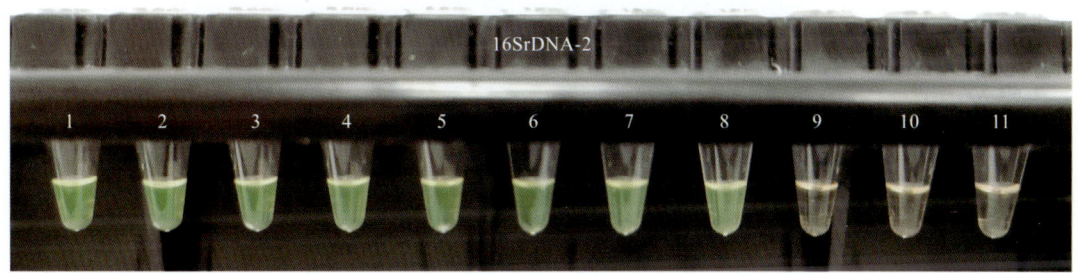

1—2 ng/μL；2—200 pg/μL；3—20 pg/μL；4—2 pg/μL；5—200 fg/μL；
6—20 fg/μL；7—2 fg/μL；8—200 ag/μL；9—20 ag/μL；10—2 ag/μL；11—去离子水。

图3-6　槟榔黄化植原体16SrDNA-2和16SrDNA-3 LAMP引物组灵敏度变色反应（Yu et al., 2020）

应用建立的槟榔黄化植原体LAMP检测体系，对采自海南保亭、屯昌、万宁等不同地区的槟榔黄化病样品进行检测（Yu et al., 2020），确定槟榔黄化植原体LAMP检测体系的稳定性。首先通过P1/P7、R16mF2/R16mR1引物PCR扩增对供试样品进行植原体检测，在保亭、屯昌、万宁的槟榔黄化病供试样品中均检测到槟榔黄化植原体，在万宁的无症槟榔供试样品中均未检测到槟榔黄化植原体。通过槟榔黄化植原体16SrDNA-2和-3 LAMP检测体系，60 min内在保亭、屯昌、万宁等海南不同地区的槟榔黄化植原体样品中均出现"S"形曲线，反应前加入显色液后反应结果均变成翠绿色，而槟榔健康样品中均未出现"S"形曲线，反应前加入显色液后反应结果均变成翠绿色。由此可知，本研究基于16S rDNA基因建立的2套槟榔黄化植原体LAMP检测引物组16SrDNA-2和16SrDNA-3，在槟榔黄化植原体快速检测应用中具有很好的稳定性，检测效率也相对较高，如图3-7、图3-8、图3-9、图3-10所示。LAMP检测技术快速高效、操作简便、结果可视化，适合于槟榔黄化病田间快速诊断、种苗带菌检测和抗性槟榔品种选育等应用研究，这对槟榔黄化病病原检测鉴定、病害扩散流行和科学防控等具有重要意义。

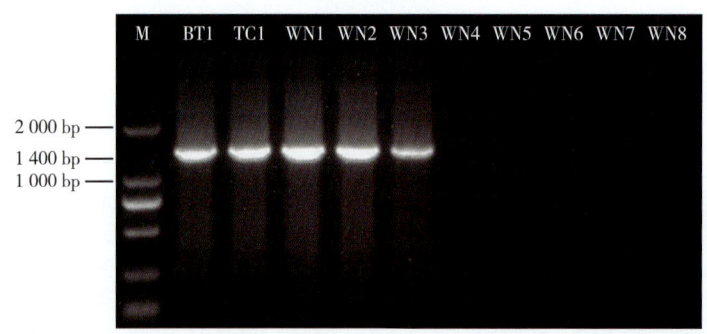

M—Marker2000分子标记；BT1—海南保亭槟榔黄化病样品；TC1—海南屯昌槟榔黄化病样品；WN1至WN3—海南万宁槟榔黄化病样品；WN4至WN8—海南万宁槟榔健康样品。

图3-7　海南不同地区槟榔黄化病和无症供试样品PCR检测（Yu et al.，2020）

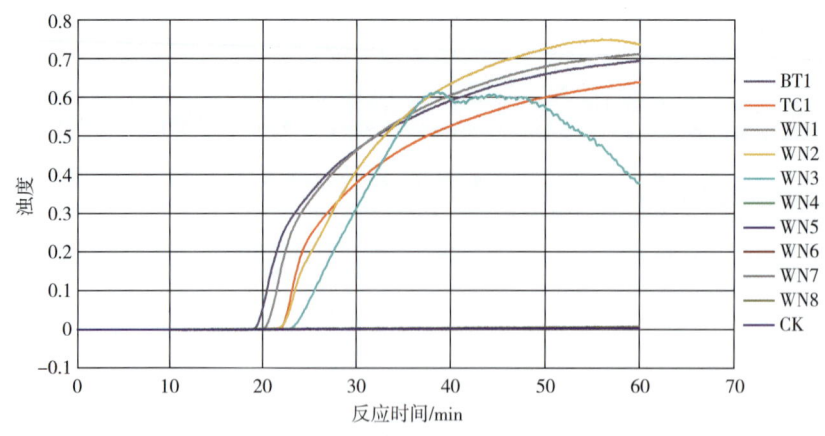

BT1—海南保亭槟榔黄化病样品；TC1—海南屯昌槟榔黄化病样品；WN1至WN3—海南万宁槟榔黄化病样品；WN4至WN8—海南万宁槟榔健康样品；CK—去离子水。

图3-8　槟榔黄化植原体16SrDNA-2 LAMP引物组稳定性实时扩增曲线（Yu et al.，2020）

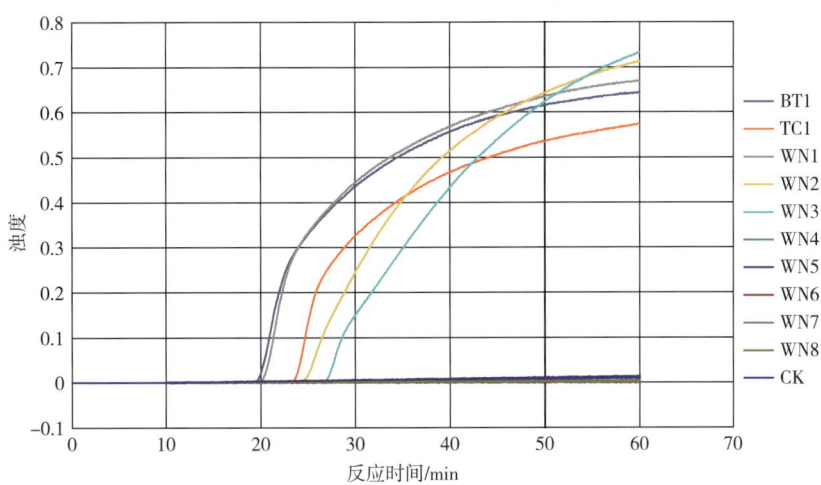

BT1—海南保亭槟榔黄化病样品；TC1—海南屯昌槟榔黄化病样样品；
WN1至WN3—海南万宁槟榔黄化病样品；WN4至WN8—海南万宁槟榔健康样品；CK—去离子水。

图3-9　槟榔黄化植原体16SrDNA-3 LAMP引物组稳定性实时扩增曲线（Yu et al.，2020）

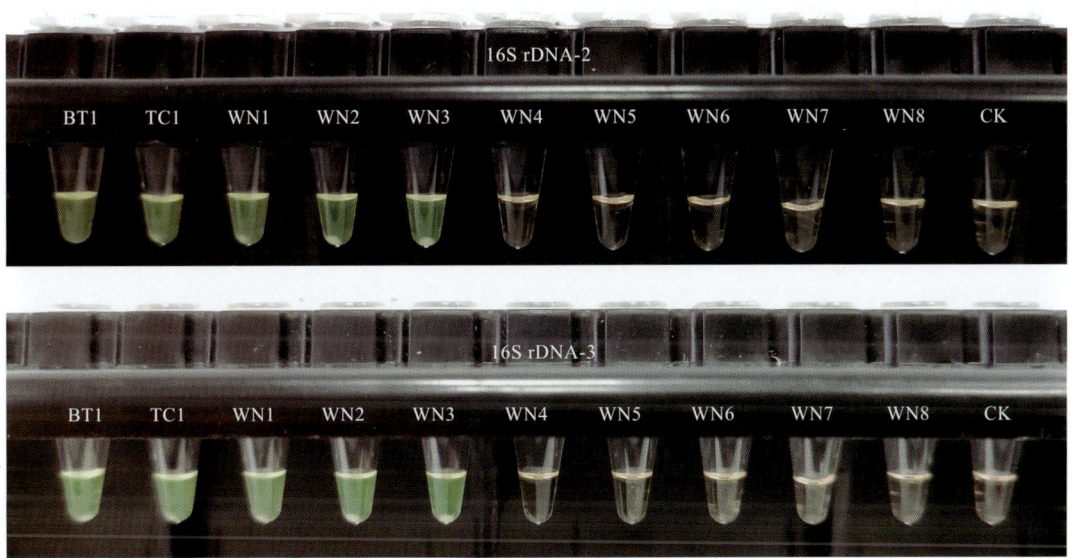

BT1—海南保亭槟榔黄化样品；TC1—海南屯昌槟榔黄化病样样品；
WN1至WN3—海南万宁槟榔黄化病样品；WN4至WN8—海南万宁槟榔健康样品；CK—去离子水。

图3-10　槟榔黄化植原体16SrDNA-2和16SrDNA-3 LAMP引物组稳定性变色反应（Yu et al.，2020）

椰子是我国华南沿海地区及海南岛主要的特色经济作物和重要的文化标志，具有重要的经济、绿化、文化和生态价值。由植原体引起的椰子致死性病害是椰子种植上的一种毁灭性病害，该病害传播快、致病性高、危害十分严重（Gurr et al.，2016）。椰子致死性黄化植原体种类丰富、地理分布广泛。地理分布上，在美洲、非洲和东南亚等地区均有分布（Gurr et al.，2016）。椰子致死性黄化相关植原体的种类，包括16Sr I

（-B）、16SrⅣ（-A、-B、-C、-E、-F）、16SrⅪ（-A、-B）、16SrⅩⅣ、16SrⅩⅫ（-A、-B）、16SrⅩⅫⅫ（-B）等多个组或亚组，不同组的椰子致死性植原体遗传变异水平较高（Bertaccini et al.，2014；Gurr et al.，2016；Babu et al.，2021）。截至目前，我国尚未见椰子致死性植原体病害的相关报道。但是已有研究表明，我国为椰子植原体的适生区，我国华南沿海及海南大部分地区为中、高风险区（朱辉等，2010；曹学仁等，2014）。华南沿海及海南大部分地区是我国椰子主产区，一旦国外椰子致死性植原体入侵，对海南乃至我国的椰子产业将是毁灭性的打击。

　　椰子致死性病原菌植原体有6组15个亚组以上，已报道的椰子致死性植原体通用型检测方法仅有巢式PCR，巢式PCR需要进行2轮PCR扩增，操作复杂，该方法反应时间较长，约为5 h，且第二轮PCR易造成污染，检测结果需要通过基因测序确定（Gurr et al.，2016；Dickinson & Hodgetts，2013）。因此，开展相关快速高效的椰子致死性植原体检测技术研究对于相关病害的检测、监测至关重要。为研发椰子致死性植原体通用型快速可视化检测技术，提高椰子致死性植原体检测效率及检测客观性，不因椰子致死性植原体种类差异而造成漏检。Yu等（2023）根据GenBank已报道的6类属于不同组的椰子致死性植原体16S rRNA基因序列的多重比对分析，获取6组椰子致死性植原体的16S rRNA基因保守区域序列。针对多种椰子致死性植原体的保守区域序列及其遗传变异特征，设计、筛选快速可视化检测引物组，建立针对不同椰子致死性植原体通用型的快速可视化检测技术。通过DNAMAN6.0对16SrⅠ组（KY814724）、16SrⅣ组（HQ613874）、16SrⅪ组（FJ794816）、16SrⅩⅣ组（EU636906）、16SrⅩⅫ组（Y14175）和16SrⅩⅫⅫ组（EU498727）进行多重比对，基于多重比对结果获得长为1 200～1 210 bp的保守区域。将6个椰子致死性植原体长为1 200～1 210 bp的16S rRNA基因保守序列人工合成后插入pUC57质粒，构建含有6组椰子致死性植原体16S rRNA靶标基因序列的重组质粒作为DNA模板，作为16SrⅠ组、16SrⅣ组、16SrⅪ组、16SrⅩⅣ组、16SrⅩⅫ组和16SrⅩⅫⅫ组椰子致死性植原体DNA模板的重组质粒分别为CO>1-1、CO>4-5、CO>11-1、CO>14-1、CO>22-1和CO>32-1。重组质粒由生工生物工程（上海）股份有限公司合成。基于NCBI可获得椰子致死性植原体16S rRNA基因的保守区域序列，利用环介导等温扩增法引物在线设计软件PrimerExplorer V4（http://primerexplorer.jp/e/）设计椰子致死性植原体的通用型快速可视化检测引物。根据环介导等温扩增引物设计原则，针对16SrⅠ组（KY814724）、16SrⅣ组（HQ613874）、16SrⅪ组（FJ794816）、16SrⅩⅣ组（EU636906）、16SrⅩⅫ组（Y14175）和16SrⅩⅫⅫ组（EU498727）椰子致死性植原体16S rRNA基因序列的共同保守区域序列，设计多套环介导等温扩增引物组，从中筛选出通用型的快速可视化检测引物组。环介导等温扩增引物组包括正向外引物F3、反向外引物B3、正向内引物FIP（FIC+F2）、反向内引物BIP（BIC+B2）。引物由生工生物工程（上海）股份有限公司按照HPLC纯度级别合成。引物序列如表3-3所示。

表3-3　环介导等温扩增引物组Co-4和Co-6序列信息表（Yu et al.，2023）

引物组	引物名称	引物序列（5'-3'）
Co-4	Co-F3-4	GGGTCTTTACTGACGCTGAG
	Co-B3-4	GGTTTTTCGGGTACCTTCGA
	Co-FIP-4	ACTCATCGTTTACGGCGTGGACGCACGAAAGCGTGGGGAG
	Co-BIP-4	TCCGCCTGAGTAGTACGTACGCCCGCTTGTGCGGAGTC
	Co-LF-4	ACCAGGGTATCTAATCCTGTTTG
Co-6	Co-F3-6	GCGTCACATTAGTTAGTTGGT
	Co-B3-6	TACTTCATCGTTCACGCG
	Co-FIP-6	ATGTGGCCGTTCAACCTCTCGGGGTAAAGGCCTACCAA
	Co-BIP-6	GACTGAGACACGGCCCAAACGGTCAGAGTTTCCTCCATT
	Co-LF-6	CCCGGCTACACATCATAGTC
	Co-LB-6	GCAGCAGTAGGGAATTTTCGG

Yu等（2023）基于椰子致死性植原体通用型快速可视化检测引物组Co-4在含有6组椰子致死性植原体16S rRNA基因的重组质粒CO>1-1、CO>4-5、CO>11-1、CO>14-1、CO>22-1、CO>32-1和环介导等温扩增试剂盒（Eiken China Co.，Ltd.）自带的阳性对照中均检测到扩增曲线，而在pUC57空载体、ddH$_2$O和环介导等温扩增试剂盒（Eiken China Co.，Ltd.）自带的阴性对照中均未检测到扩增曲线。6组椰子致死性植原体DNA模板的扩增曲线均在40 min内出现，如图3-11所示。基于椰子致死性植原体通用型快速可视化检测引物组Co-6在含有6组椰子致死性植原体16S rRNA基因的重组质粒CO>1-1、CO>4-5、CO>11-1、CO>14-1、CO>22-1、CO>32-1和环介导等温扩增试剂盒（Eiken China Co.，Ltd.）自带的阳性对照中均检测到扩增曲线，而在pUC57空载体、ddH$_2$O和环介导等温扩增试剂盒（Eiken China Co.，Ltd.）自带的阴性对照中均未检测到扩增曲线。6组椰子致死性植原体DNA模板的扩增曲线均在30 min内出现，如图3-12所示。在环介导等温扩增反应体系中预先加入荧光目视检测试剂FD显色剂（钙黄绿素-氯化锰溶液），加入椰子致死性植原体通用型快速可视化检测引物组Co-4和Co-6后64℃恒温扩增60 min，6组椰子致死性植原体DNA模板CO>1-1、CO>4-5、CO>11-1、CO>14-1、CO>22-1、CO>32-1和环介导等温扩增试剂盒（Eiken China Co.，Ltd.）自带的阳性对照的反应体系颜色均变为翠绿色，而pUC57空载体、ddH$_2$O和环介导等温扩增试剂盒（Eiken China Co.，Ltd.）自带的阴性对照的反应体系颜色均未变，仍为橘黄色，如图3-11、图3-12所示。6组椰子致死性植原体DNA模板快速可视化检测的显色反应与其环介导等温扩增检测的扩增曲线结果一致。

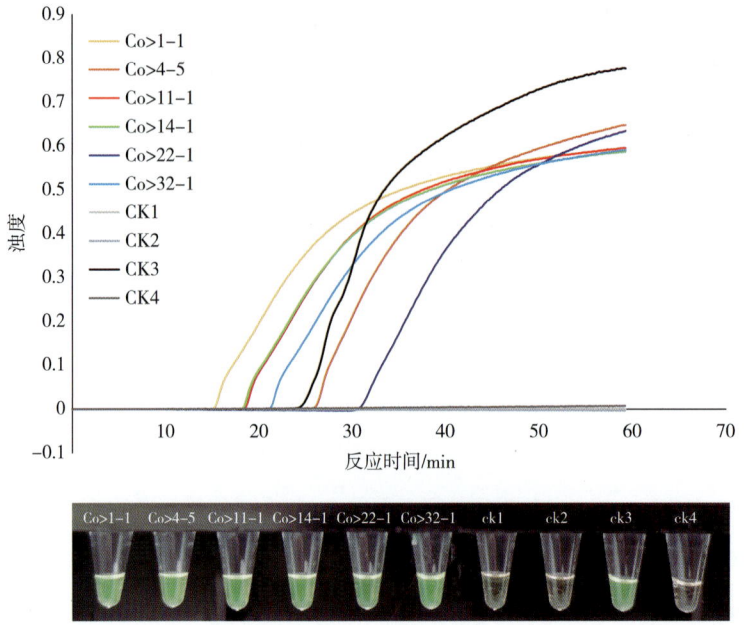

Co>1-1、Co>4-5、Co>11-1、Co>14-1、Co>22-1、Co>32-1—不同植原体组的重组质粒；
CK1—空载体pUC57；CK2—ddH$_2$O；CK3—环介导等温扩增试剂盒（Eiken China Co., Ltd.）
自带的阳性对照；CK4为环介导等温扩增试剂盒自带的阴性对照。

图3-11　椰子致死性植原体快速可视化检测引物组Co-4扩增曲线与显色反应（Yu et al., 2023）

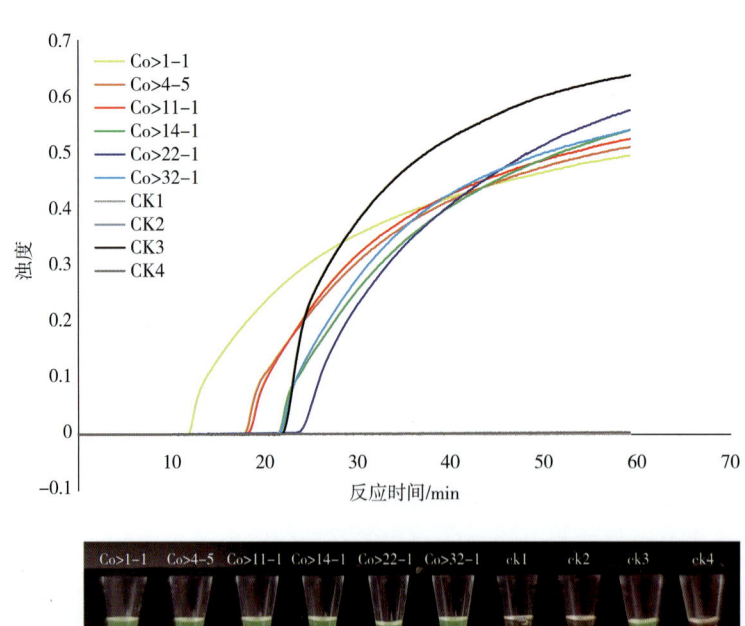

Co>1-1、Co>4-5、Co>11-1、Co>14-1、Co>22-1、Co>32-1—不同植原体组的重组质粒；
CK1—空载体pUC57；CK2—ddH$_2$O；CK3—环介导等温扩增试剂盒（Eiken China Co., Ltd.）
自带的阳性对照；CK4—环介导等温扩增试剂盒自带的阴性对照。

图3-12　椰子致死性植原体快速可视化检测引物组Co-6扩增曲线与显色反应（Yu et al., 2023）

Yu等（2023）基于6组椰子致死性植原体16S rRNA基因重组质粒梯度浓度，对椰子致死性植原体通用型快速可视化检测引物组Co-4灵敏度进行分析，结果表明，6组椰子致死性植原体DNA模板CO>1-1、CO>4-5、CO>11-1、CO>14-1、CO>22-1、CO>32-1最低检测限浓度不同，最低检测限范围在0.1 ag/μL ~ 1 pg/μL。通用型快速可视化检测引物组Co-4检测16Sr XIV组椰子致死性植原体灵敏度最高，为0.1 ag/μL；检测16Sr XXII组椰子致死性植原体灵敏度最低，为1 pg/μL；检测16SrXI组和16Sr XXXII组椰子致死性植原体灵敏度相同，均为1 ag/μL；检测16Sr I组和16SrIV组椰子致死性植原体灵敏度分别为10 ag/μL和100 fg/μL。由此可知，相同快速可视化检测引物组，针对不同椰子致死性植原体DNA模板最低检测限浓度差异较大，基于椰子致死性植原体通用型快速可视化检测引物组Co-4，6组椰子致死性植原体之间灵敏度差异为10^7倍，如图3-13所示。

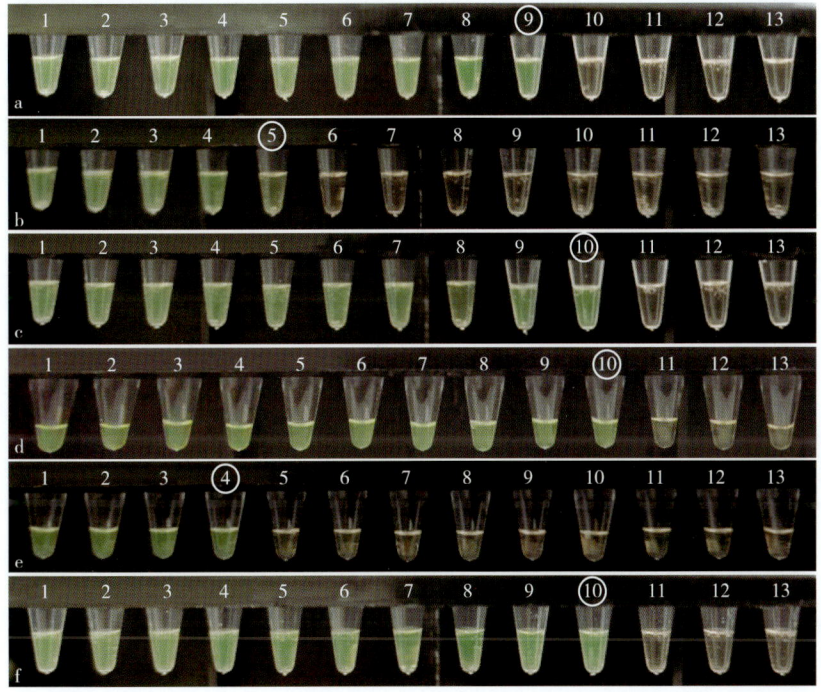

a、b、c、d、e和f—Co>1-1、Co>4-5、Co>11-1、Co>14-1、Co>22-1、Co>32-1重组质粒；4.8×10^{10}（1）、4.8×10^9（2）、4.8×10^8（3）、4.8×10^7（4）、4.8×10^6（5）、4.8×10^5（6）、4.8×10^4（7）、4.8×10^3（8）、4.8×10^2（9）、4.8×10^1（10）、4.8（11）、4.8×10^{-1}（12）拷贝/200 μL，ddH_2O（13）。

图3-13 椰子致死性黄化植原体通用型引物组Co-4灵敏度显色图（Yu et al.，2023）

Yu等（2023）基于6组椰子致死性植原体16S rRNA基因重组质粒梯度浓度，对椰子致死性植原体通用型快速可视化检测引物组Co-6灵敏度进行分析，结果表明，6组椰子致死性植原体DNA模板CO>1-1、CO>4-5、CO>11-1、CO>14-1、CO>22-1、CO>32-1最低检测限浓度不同，最低检测限范围在1 ag/μL ~ 1 pg/μL。通用型快速可视化检测引物组Co-6检测16Sr XIV组椰子致死性植原体灵敏度最高，为1 ag/μL；检测16Sr XXII组椰子致死性植原体灵敏度最低，为1 pg/μL；检测16SrIV组、16SrXI组和16Sr XXXII组椰子致死性植原体灵敏度相同，均为1 fg/μL；检测16SrⅠ组椰子致死性植原体灵敏度分别为100 ag/μL。由此可知，基于椰子致死性植原体通用型快速可视化检测引物组Co-6，6组椰子致死性植原体之间灵敏度差异亦是较大，为10^6倍，如图3-14所示。建立的针对16SrⅠ、16SrⅣ、16SrⅪ、16SrXIV、16SrXXII、16SrXXXII共6组遗传变异水平较高的椰子致死性植原体通用型快速可视化检测方法，对于椰子致死性植原体及其病害的检测鉴定、诊断预警、流行监测等具有重要的理论意义和应用价值。

a、b、c、d、e和f—Co>1-1、Co>4-5、Co>11-1、Co>14-1、Co>22-1、Co>32-1重组质粒；$4.8×10^{10}$（1）、$4.8×10^9$（2）、$4.8×10^8$（3）、$4.8×10^7$（4）、$4.8×10^6$（5）、$4.8×10^5$（6）、$4.8×10^4$（7）、$4.8×10^3$（8）、$4.8×10^2$（9）、$4.8×10^1$（10）、4.8（11）、$4.8×10^{-1}$（12）拷贝/200 μL，ddH_2O（13）。

图3-14　椰子致死性黄化植原体通用型引物组Co-6灵敏度显色图（Yu et al.，2023）

七、微滴式数字PCR扩增技术

微滴式数字PCR（droplet digital PCR，ddPCR）采用一种全新的方式进行核酸分子的定性定量分析，即在传统的PCR扩增前对样品进行微滴化处理，将含有核酸分子的荧光PCR反应体系"分割"成数万个纳米级的微滴，经PCR扩增后，对每个微滴进行检测，有荧光信号的微滴判读为1，没有荧光信号的微滴判读为0，从而将荧光模拟信号数字化，因此该技术被称为"数字PCR"（www.bio-rad.com；Day et al.，2013）。最终根据泊松分布原理及阳性微滴的个数与比例，分析软件即可给出待检靶标分子的起始拷贝数浓度（www.bio-rad.com；Day et al.，2013）。与传统PCR、定量PCR相比，ddPCR检测结果的精确度、准确性和灵敏度更佳。其定量的结果不再依赖于Cq值而直接给出靶标序列的起始拷贝数浓度，实现真正意义上的绝对定量（Day et al.，2013）。槟榔黄化植原体在发病槟榔组织中含量较低，且浓度分布不均（Yu et al.，2021）。因此，为了精确、灵敏地检测鉴定待检槟榔组织中植原体的有无及分布情况，建立槟榔黄化植原体高灵敏度检测技术势在必行。

Yu等（2022）基于16SrⅠ-B亚组植原体*tuf*基因序列（序列号：AP006628），设计槟榔黄化植原体ddPCR检测引物对和探针。引物对Atf/Atr与探针AtProbe的序列信息如下所示。

Atf：5′AAATTTATACGAAACAAACCGCATTTA3′；

Atr：5′ACGTGAGCATAGTGACGTTTTTC3′；

AtProbe：5′AGCAGCTATTACCCAAGTTTTGTCTAC3′。

探针AtProbe的5′端由VIC修饰，3′端由BHQ1修饰，引物与探针由生工生物工程（上海）股份有限公司合成并经HPLC级别纯化。

将总体积为24 μL ddPCR反应体系放入QX200微滴生成仪中生成微滴，将生成微滴后的反应体系放入T100型梯度PCR仪中进行扩增，扩增结束后将PCR产物轻轻平衡放入微滴读取仪中，通过QuantaSoft软件读取数据，每份样品重复检测3次。ddPCR反应体系总体积为24 μL，反应体系包括12 μL ddPCR预混液（无dUTP），900 nmol/L引物Atf/Atr10 μmol/L，250 nmol/L探针AtProbe10 μmol/L，DNA模板1 μL，加ddH$_2$O补齐至24 μL。ddPCR反应条件为95℃10 min；94℃ 30 s，53℃ 60 s，共40个循环；98℃10 min。

Yu等（2022）以我国海南槟榔黄化植原体为研究对象，建立槟榔黄化植原体ddPCR检测体系。基于槟榔黄化病样品AYL1、AYL2、AYL3及健康槟榔样品H1、H2、H3，建立槟榔黄化植原体ddPCR检测体系。由结果可知，通过引物Atf/Atr和探针AtProbe可在槟榔黄化病样品中检测到明显的槟榔黄化植原体信号（图3-15）。如图3-15所示，A02、B02、C02代表的槟榔黄化病样品AYL1、AYL2、AYL3的阳性微滴与阴性微滴分离明显，有明显的阳性结果；D02、E02、F02代表的健康槟榔样品H1、H2、H3无阳性结果；G02为DNA提取过程洗脱DNA用的TE缓冲液，H02为ddPCR反应体系中加入的ddH$_2$O，TE缓冲液和ddH$_2$O对照中均无阳性结果。绝对定量结果显示，样品AYL1、AYL2、AYL3的植原体*tuf*基因拷贝数浓度分别为6.9拷贝/μL、6.0拷贝/μL、6.0拷贝/μL（图3-16A）。由图3-16B可知，每个反应的微滴生成数均超过10 000。因此，基于*tuf*基因设计的引物Atf/Atr和探针AtProbe对槟榔黄化植原体的定量结果可信，建立的槟榔黄化植原体ddPCR检测体系检测效果较好。Yu等（2022）采用建立的槟榔黄化植原体ddPCR检测技术对采自海南不同地区

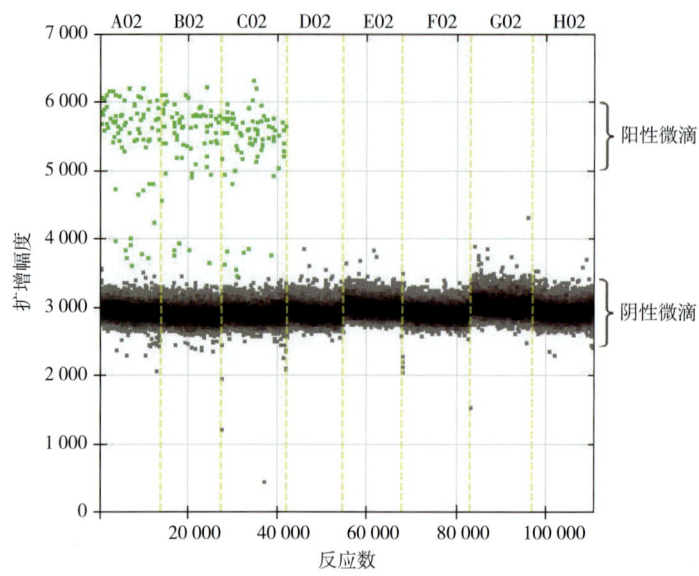

A02、B02、C02柱—槟榔黄化病病样AYL1、AYL2、AYL3；D02、E02、F02—无症状槟榔样品H1、H2、H3；G02柱—用于DNA提取的TE缓冲液；H02柱—用于ddPCR反应的ddH$_2$O。

图3-15　通过引物Atf/Atr和探针AtProbe检测槟榔黄化植原体（Yu et al.，2022）

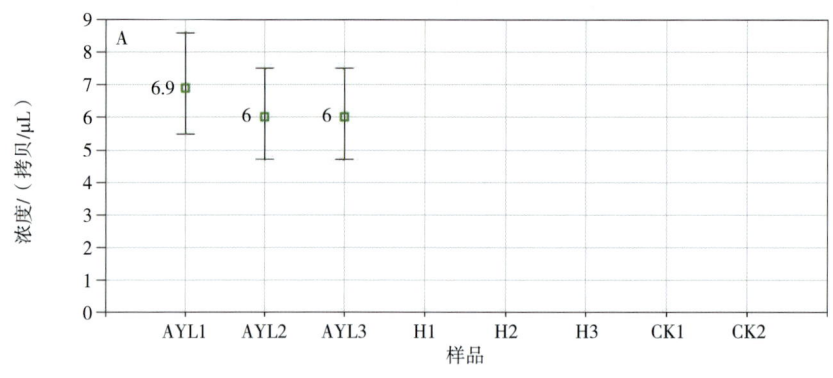

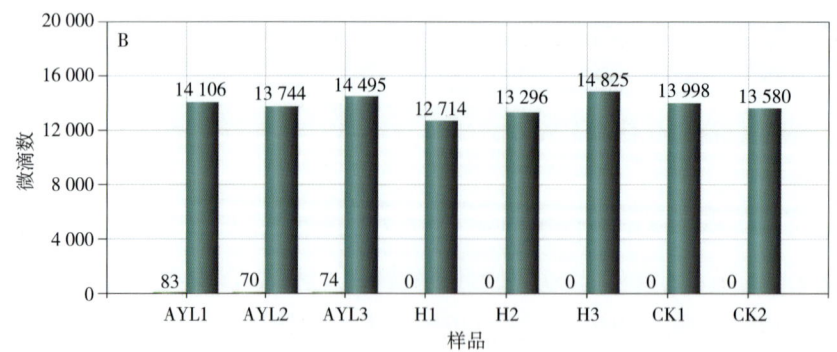

AYL1、AYL2、AYL3—槟榔黄化病病样；H1、H2、H3—无症状槟榔样品；
CK1—用于DNA提取的TE缓冲液；CK2—代表用于ddPCR反应的ddH$_2$O。

图3-16　通过引物Atf/Atr和探针AtProbe对槟榔黄化植原体进行绝对定量检测（Yu et al.，2022）

槟榔黄化病样品进行植原体检测，ddPCR检测技术在采自不同地区的槟榔黄化病样品中均可检测到槟榔黄化植原体，检测结果可信（图3-17）。如图3-17所示，屯昌槟榔黄化病样品AYL4、AYL5的植原体浓度分别为0.12拷贝/μL、0.22拷贝/μL；采自文昌的样品AYL6、AYL7、AYL8的植原体浓度分别为0.07拷贝/μL、0.07拷贝/μL、0.16拷贝/μL；万宁样品AYL9、AYL10的植原体浓度分别为0.23拷贝/μL、0.09拷贝/μL；CK为ddH$_2$O对照，无植原体检测信号。在检测到植原体的不同槟榔黄化病样品中，植原体含量最低可达到0.07拷贝/μL。

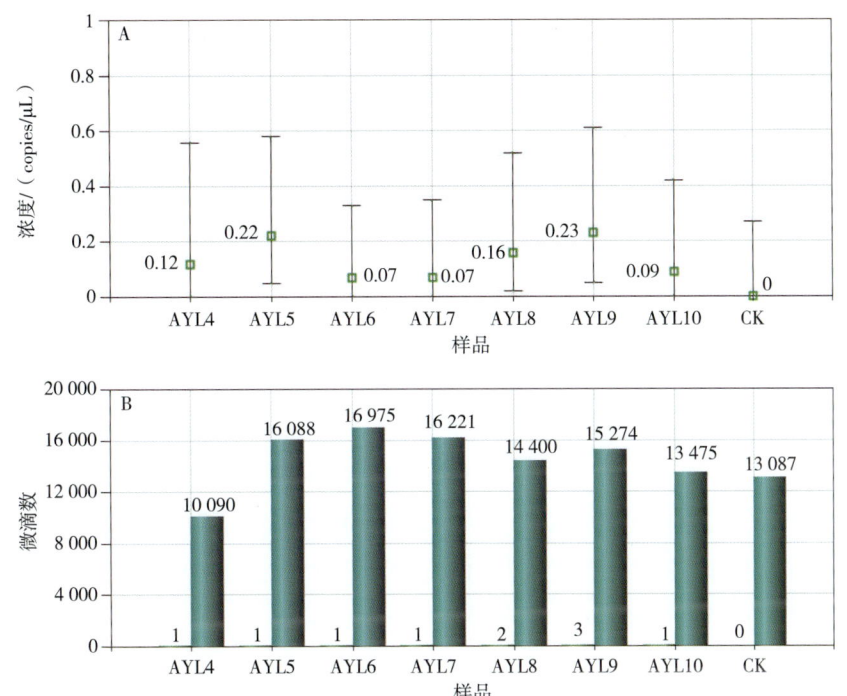

AYL4、AYL5—采自海南屯昌的槟榔黄化病病样；AYL6、AYL7、AYL8—采自海南文昌的病样；AYL9、AYL10—采自海南万宁的病样；CK—用于作为参照的ddH$_2$O。

图3-17 槟榔黄化植原体绝对定量检测（Yu et al.，2022）

根据ddPCR操作要求（www.bio-rad.com），ddPCR检测体系中加入的DNA模板溶液中的靶标分子拷贝数应在1万~10万个，ddPCR运行过程中微滴生成仪会吸取20 μL反应体系用于微滴生成。因此，ddPCR可以达到的靶标分子浓度检测限为0.05拷贝/μL，即5.0×10^{-2}拷贝/μL。研究表明，槟榔黄化植原体LAMP检测技术能达到的DNA模板浓度检测限为5.3×10^{1}拷贝/μL（Yu et al.，2020）；荧光定量PCR检测扩增灵敏度能达到的DNA模板浓度为3.17×10^{2}拷贝/μL（任争光等，2015）；巢式PCR检测扩增灵敏度能达到的DNA模板浓度的为3.17×10^{4}拷贝/μL（任争光等，2015）。由此可知，ddPCR检测技术最为灵敏，与LAMP检测技术相比检测灵敏度提高了约1 000倍。Yu等（2022）检测到的槟榔黄化植原体最低浓度为0.07拷贝/μL，这一浓度已超出LAMP、荧光定量PCR和巢式PCR等检测技术可检测的DNA模板浓度范围。Yu等（2022）建立的槟榔黄化植原体ddPCR检测技术在植原体病害监测、槟榔种苗检疫及槟榔黄化病综合防治等方面应用前景较为广泛，对于槟榔黄化病病原的定量检测、病害的流行监测及其传播昆虫媒介、中间寄主的筛选鉴定等研究具有较为重要的意义，为我国海南岛槟榔黄化病及其他植原体病害的流行监测和防控管理提供理论参考和技术支撑。

八、微阵列生物芯片检测技术

基因芯片又称DNA芯片或DNA微阵列，是指固着在载体上的高密度DNA微点阵。该技术是根据核酸分子杂交原理衍生而来，将序列已知的核酸作为探针，对序列未知的核酸进行杂交检测（Campas et al.，2004）。根据载体类型以及用途的不同，基因芯片可以分为不同的类型，如基因表达芯片、测序芯片、诊断芯片等。相比于传统的核酸杂交技术，基因芯片具有快速、平行、高效、高通量、高灵敏度、可自动化操作等特点，自被发明以来，在医学、生命科学、农业、林业及环境科学等领域得到了广泛的应用（Barba et al.，2008；Myles et al.，2015；Yip et al.，2015；王圣洁等，2017a）。王圣洁等（2017a）建立了管芯片检测和鉴别植原体技术体系。

参考Nicolaise等（2017）针对植原体的16Sr DNA设计的探针组，包含一对通用引物Pp-fwd（5′-AGTGGCGAACGGGTGAGTAAC-3′）和Pp-rev（5′-CGTTTACGGCGTGGACTACCAG-3′）以及一组针对不同组植原体特异性的探针（表3-4）。将人工合成的探针序列，经点样机按设计的排布点在管芯片中。探针序列及阳性对照在载体上的分布情况如图3-18。

表3-4 探针序列（王圣洁等，2017a）

探针编号	探针序列（5′-3′）	GenBank登录号
PpⅠ-465	TATTAGGGAAGAATAAATGATGGAAAAATC	AF222064
PpⅡ-471	CGAACCATTTGTTTGCCGGTA	AF028813
PpⅡ-629	CGTTGTCCGGCTATTGAAACTGC	—
PpⅢ-478	GTGGAAAAACTCCCTTGACGGTACTTAAT	AF173558
PpⅣ-630	CTTAACGTTGTCCTGCTAGAGAAACTGTT	AF361020
PpⅤ-221	AGACCTTCTTCGGAGGGTATGCTTAA	AY072722
PpⅥ-276	TTAGTTGGTAGAGTAAAAGCCTACCAAGAC	AF190224
PpⅦ-621	TATAGAAACTACCTTGACTAGAGTTAGATAGAG	AF105315
PpⅧ-634	ACGCTTAACGTTGTTTTGTTATAGAAACTG	AF248956
PpⅨ-609	AACGCTGTAsCGCTATAGAAACTGTCTG	AF515637

（续表）

探针编号	探针序列（5′-3′）	GenBank登录号
PpⅩ-183	GGATAGGAAGTTTTAAGGCATCTTGAAAC	AF248958
PpⅩ-224	AGGGTATGCTAAGAGATGGGCTTGC	—
PpⅫ-465	GGGAAGAAAAGATGGTGGAAAAACC	AF248959
PpⅫ-480	GTGGAAAAACCATTATGACGGTACCT	
PpⅩⅣ-585	GAAACTATCAGACTAGAGTGAGATAGAGGCAAG	AF509321
Pp-148	TTTCGGCAATGGAGGAAACTCTGAC	AY741532
Pp-502	AGGCGGCTyrCTGGGTCTTTACT	—

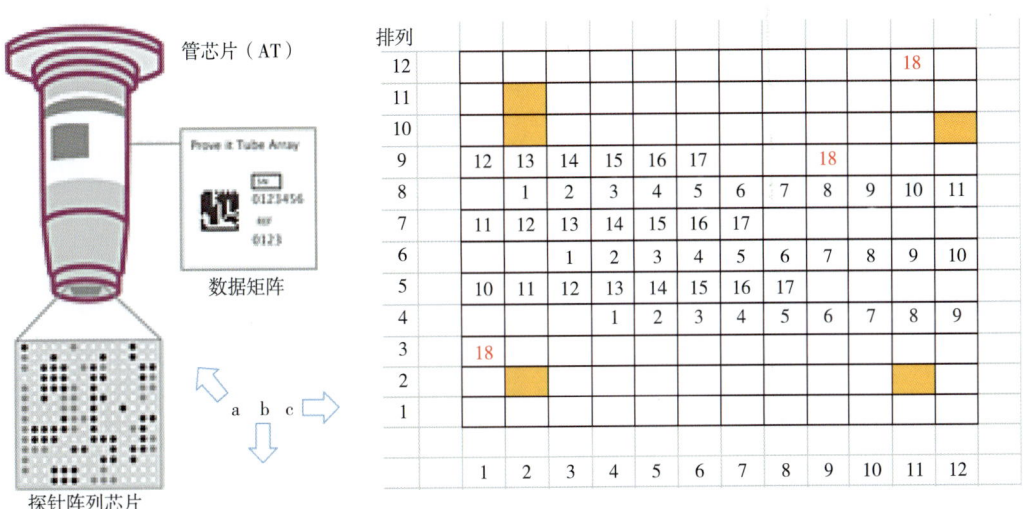

序号	探针（5′-3′）	序号	探针（5′-3′）
1	PpⅠ-465（16SrⅠ组）	10	PpⅨ-609（16SrⅨ组）
2	PpⅡ-471（16SrⅡ组）	11	PpⅩ-183（16SrⅩ组）
3	PpⅡ-629（16SrⅡ组）	12	PpⅩ-224（16SrⅩ组）
4	PpⅢ-478（16SrⅢ组）	13	PpⅫ-465（16SrⅫ组）
5	PpⅣ-630（16SrⅣ组）	14	PpⅫ-480（16SrⅫ组）
6	PpⅤ-221（16SrⅤ组）	15	PpⅩⅣ-585（16SrⅩⅣ组）
7	PpⅥ-276（16SrⅥ组）	16	Pp-148（16Sr ALL）
8	PpⅦ-621（16SrⅦ组）	17	Pp-502（16SrALL）
9	PpⅦ-634（16SrⅦ组）	18	Biotin-Marke_2,5μM

a—管芯片示意图；b—探针分布图；c—探针点代号。

图3-18 管芯片示意及探针分布图（王圣洁等，2017a）

王圣洁等（2017a）研究结果表明，15种病害样品中，13种获得显著的阳性杂交信号，并且所有的健康对照都呈现为阴性。13种植原体病害根据16Sr DNA直接测序可分为16SrⅠ、16SrⅡ、16SrⅢ、16SrⅩⅨ共4组植原体。在所有探针中，植原体的通用探针（Pp-502）可以检测到所有确定的植原体样品。16SrⅠ组特异性探针（PpⅠ-465）可以确定16SrⅠ组的泡桐丛枝、苦楝丛枝、桑树萎缩和莴苣黄化4种植原体样品。16SrⅡ组特异性探针（PpⅡ-629）仅可以确定16SrⅡ组的花生丛枝、甘薯丛枝和臭矢菜丛枝3种植原体样品。但16SrⅤ组的枣疯病、樱桃致死黄化和重阳木丛枝及16SrⅩⅨ组的板栗黄化皱缩植原体与其他组专化性探针皆有明显的交叉杂交信号，如图3-19所示。

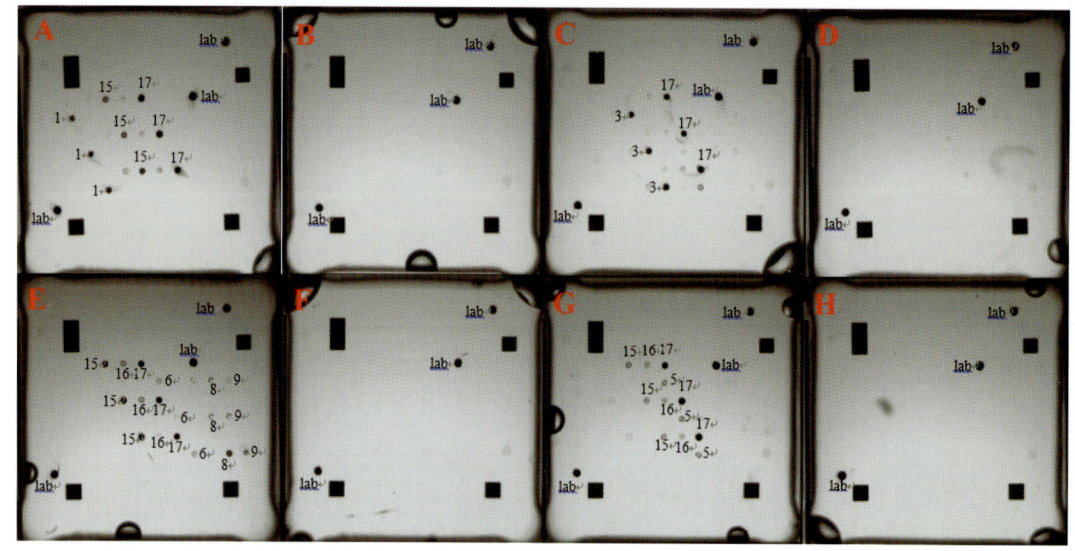

A—泡桐丛枝；B—健康泡桐；C—花生丛枝；D—健康花生；
E—枣疯病；F—健康枣树；G—板栗皱缩黄化；H—健康板栗。

图3-19 芯片杂交结果（王圣洁等，2017a）

相比PCR扩增的凝胶电泳检测，管芯片检测的灵敏度提高了1 000倍。将泡桐丛枝样品PaWB-HBBD的16Sr RNA基因扩增后，连接到克隆载体pMD18-T上后组建重组质粒，提取并纯化后进行10倍梯度稀释，并作为模板分别进行PCR检测和管芯片检测。实验结果显示，以植原体通用引物R16mF2/R16mR1的传统PCR检测，在质粒浓度稀释到10^{-4}倍时，达到凝胶电泳的检测限度（图3-20）。但是管芯片检测在质粒稀释到10^{-7}倍时，仍可出现较好的杂交信号（图3-21）。因此可以表明相比于传统PCR检测，管芯片检测植原体的方法的灵敏度提高了1 000倍。采用管芯片检测的方法，不仅能实现灵敏度的提高，还能实现植原体不同组的鉴定。当稀释度达到10^{-7}时，与16SrⅩⅣ组探针的交叉杂交信号也明显减弱，而16SrⅠ专化性探针和通用探针的信号仍较强。对疑似植原体病害的诊断结果显示，河南濮阳的红花槐丛枝的病原应为16SrⅤ组植原体，福建福州的长春花黄化丛枝应为

16SrⅠ组植原体；而北京戒台寺牡丹黄化皱叶和内蒙古包头柳树丛枝未出现任何植原体专化的杂交信号（图3-22）。

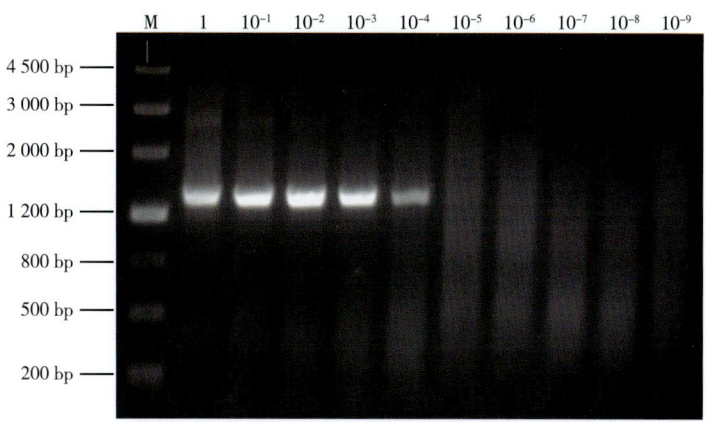

图3-20　PCR检测灵敏度（王圣洁等，2017a）

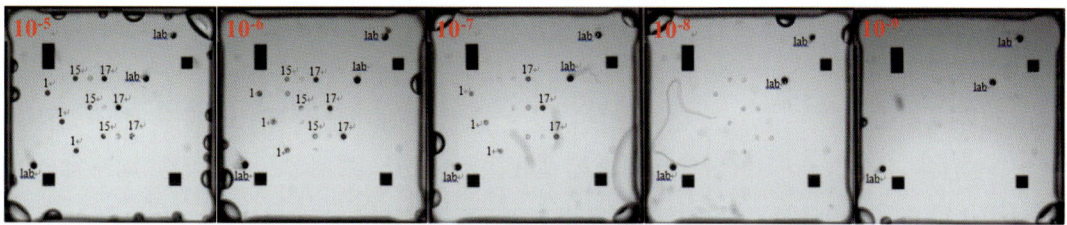

图3-21　管芯片检测灵敏度（王圣洁等，2017a）

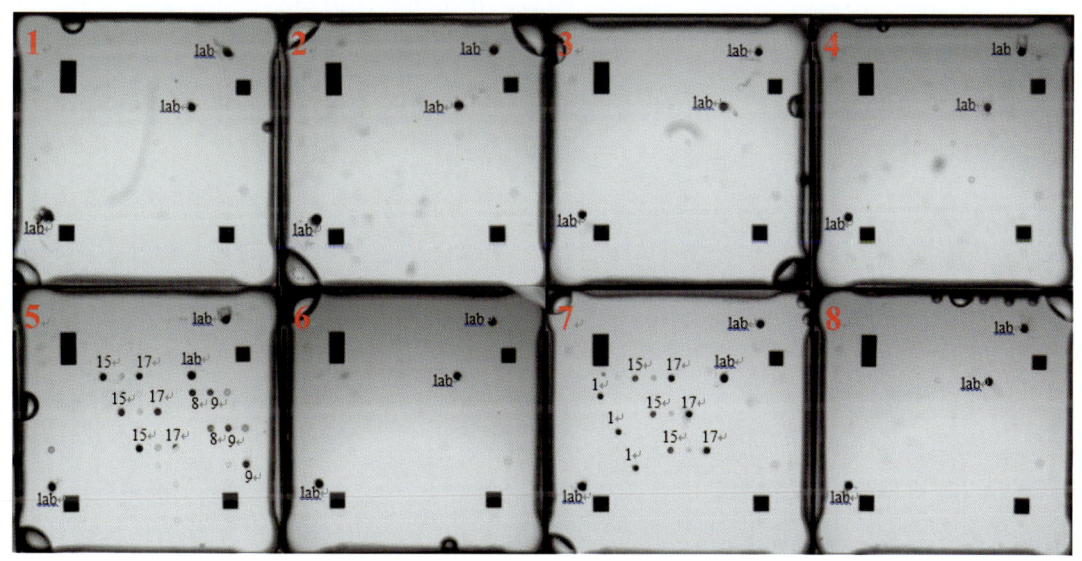

1—牡丹黄化皱缩；2—健康牡丹；3—柳树丛枝；4—健康柳树；
5—红花槐丛枝；6—健康槐树；7—长春花黄化丛枝；8—健康长春花。

图3-22　疑似植原体病害管芯片杂交结果（王圣洁等，2017a）

九、多位点检测技术

多位点序列分析（MLSA）适用于遗传关系较近的微生物遗传变异和系统发育关系研究，该方法应用广泛，能揭示微生物更多的遗传变异特征（Bishop et al., 2009; Gevers et al., 2005; Maiden et al., 1998; Saux, 2013; Spratt, 1999; Yuan et al., 2010）。目前MLSA技术也被应用于植原体遗传分析中（Arnaud et al., 2007; Danet et al., 2007; Li et al., 2014b; Music et al., 2011）。于少帅等（2017）将 *rp*、*tuf*、*secA*、*secY*、*ipt*、*dnaK*、*fusA*、*gyrB*、*pyrG*、*rpoB* 及其对应氨基酸序列（除了RP蛋白序列）共10个基因串联在一起，对我国16SrⅠ组几种主要的植原体病害进行多位点序列分析，核苷酸串联序列长为8 939～9 010 bp。苦楝丛枝病、莴苣黄化病、桑萎缩病、泡桐丛枝病和长春花绿变是中国和其他亚洲国家常见的5种对经济或观赏植物具有毁灭性损害的植物病害，基于16S rDNA序列分析引起这些植原体病害的植原体均为16SrⅠ组植原体（Tian & Raychaudhuri, 1996; Lin et al., 2014; 卢全有, 2010; 赖帆等, 2008）。苦楝丛枝病是我国华南地区常见的植原体病害，其症状主要表现为苦楝叶片减少、黄化、短枝、簇生且逐渐死亡（金开璇等, 1982）。莴苣黄化病是近期在福建三明地区发现的由植原体引起的莴苣叶片黄化症状的病害，此病害对当地莴苣栽培危害严重（林积秀等, 2013; Lin et al., 2014）。桑萎缩在我国华南地区和日本、韩国等地均有发现，并对当地的蚕丝业造成较大危害（蒯元璋, 2012; 李章宝, 1992; 卢全有, 2010）。泡桐丛枝病是泡桐生长的一种毁灭性病害，对我国从南到北的泡桐均有较大影响（Tian & Raychaudhuri, 1996; 王洁等, 2010）。16SrⅠ组植原体引起的植物病害种类多，且在世界范围内分布广泛（Lee et al., 2004）。基于单个基因及其编码蛋白序列（株系及基因信息如表3-5所示），很难将相同组的植原体株系或相同寄主的植原体株系进行区分（图3-23、图3-24、图3-25）（于少帅, 2016）。但基于MLSA序列不仅能清晰区分16SrⅠ组不同寄主的植原体株系，还能区分可将来自不同地区的相同寄主的植原体株系（图3-26、图3-27）（于少帅, 2016; Yu et al., 2017）。如来自我国不同地区的10个苦楝丛枝植原体（CWB）株系清晰地聚于4个小的分枝，即由江苏植原体株系构成的分枝1（CWB cluster 1）、由江西和湖南株系构建的分枝2（CWB cluster 2）、由福建和广州植原体株系构成的分枝3（CWB cluster 3）和由2个海南岛植原体株系构成的分枝4（CWB cluster 4）（于少帅, 2016; Yu et al., 2017）。不同地区CWB植原体株系与各自地理分布的关系示意图如图3-28所示。基于氨基酸序列的MLSA分析构建的系统发育树表明，其拓扑结构与基于核苷酸MLSA构建的系统发育的拓扑结构基因一致，只是在分枝长度所代表的遗传距离有所差异（图3-29）（于少帅, 2016; Yu et al., 2017）。

表3-5 涉及的植原体株系信息（干少帅，2016）

株系	引起的病害	采集地	级别	基因GenBank序列号									
				16S rDNA	rp	tuf	secA	secY	ipt	dnaK	fusA	gyrB	pyrG	rpoB
CWB-hnsy1	chinaberry witches'-broom	Sanya, Hainan, China	16Sr I	KP662119	KP662137	KP662155	KP662173	KP662191	KP662209	KP662227	KP662245	KP662263	KP662281	KP662299
CWB-hnsy2	chinaberry witches'-broom	Sanya, Hainan, China	16Sr I	KP662120	KP662138	KP662156	KP662174	KP662192	KP662210	KP662228	KP662246	KP662264	KP662282	KP662300
CWB-fjfz1	chinaberry witches'-broom	Fuzhou, Fujian, China	16Sr I	KP662121	KP662139	KP662157	KP662175	KP662193	KP662211	KP662229	KP662247	KP662265	KP662283	KP662301
CWB-fjfz2	chinaberry witches'-broom	Fuzhou, Fujian, China	16Sr I	KP662122	KP662140	KP662158	KP662176	KP662194	KP662212	KP662230	KP662248	KP662266	KP662284	KP662302
CWB-fjfq	chinaberry witches'-broom	Fuqing, Fujian, China	16Sr I	KP662123	KP662141	KP662159	KP662177	KP662195	KP662213	KP662231	KP662249	KP662267	KP662285	KP662303
CWB-fjya	chinaberry witches'-broom	Yong'an, Fujian, China	16Sr I	KP662124	KP662142	KP662160	KP662178	KP662196	KP662214	KP662232	KP662250	KP662268	KP662286	KP662304
CWB-jsnj	chinaberry witches'-broom	Nanjing, Jiangsu, China	16Sr I	KP662125	KP662143	KP662161	KP662179	KP662197	KP662215	KP662233	KP662251	KP662269	KP662287	KP662305
CWB-jxnc	chinaberry witches'-broom	Nanchang, Jiangxi, China	16Sr I	KP662126	KP662144	KP662162	KP662180	KP662198	KP662216	KP662234	KP662252	KP662270	KP662288	KP662306
CWB-hncs	chinaberry witches'-broom	Changsha, Hunan, China	16Sr I	KP662127	KP662145	KP662163	KP662181	KP662199	KP662217	KP662235	KP662253	KP662271	KP662289	KP662307
CWB-gdgz	chinaberry witches'-broom	Guangzhou, Guangdong, China	16Sr I	KP662128	KP662146	KP662164	KP662182	KP662200	KP662218	KP662236	KP662254	KP662272	KP662290	KP662308
LY-fjsm1	lettuce yellow	Sanming, Fujian, China	16Sr I	KP662129	KP662147	KP662165	KP662183	KP662201	KP662219	KP662237	KP662255	KP662273	KP662291	KP662309
LY-fjsm2	lettuce yellow	Sanming, Fujian, China	16Sr I	KP662130	KP662148	KP662166	KP662184	KP662202	KP662220	KP662238	KP662256	KP662274	KP662292	KP662310

（续表）

株系	引起的病害	采集地	级别	基因GenBank序列号									
				16S rDNA	rp	tuf	secA	secY	ipt	dnaK	fusA	gyrB	pyrG	rpoB
MD-ahhf	mulberry dwarf	Hefei, Anhui, China	16Sr I	KP662131	KP662149	KP662167	KP662185	KP662203	KP662221	KP662239	KP662257	KP662275	KP662293	KP662311
MD-zjca	mulberry dwarf	Chun'an, Zhejiang, China	16Sr I	KP662132	KP662150	KP662168	KP662186	KP662204	KP662222	KP662240	KP662258	KP662276	KP662294	KP662312
PaWB-sdyz	paulownia witches'-broom	Yanzhou, Shandong, China	16Sr I	KP662133	KP662151	KP662169	KP662187	KP662205	KP662223	KP662241	KP662259	KP662277	KP662295	KP662313
PaWB-jsnj	paulownia witches'-broom	Nanjing, Jiangsu, China	16Sr I	KP662134	KP662152	KP662170	KP662188	KP662206	KP662224	KP662242	KP662260	KP662278	KP662296	KP662314
PaWB-bjhr	paulownia witches'-broom	Huairou, Beijing, China	16Sr I	KP662135	KP662153	KP662171	KP662189	KP662207	KP662225	KP662243	KP662261	KP662279	KP662297	KP662315
PeV-hnhk	periwinkle virescence	Haikou, Hainan, China	16Sr I	KP662136	KP662154	KP662172	KP662190	KP662208	KP662226	KP662244	KP662262	KP662280	KP662298	KP662316
OY-M	onion yellow	Japan	16Sr I	AP006628	AP006628	AP006628	AP006628	AP006628	AP006628	AP006628	AP006628	AP006628	AP006628	AP006628
AYWB	aster yellow witches'-broom	U.S.A.	16Sr I	CP000061	CP000061	CP000061	CP000061	CP000061	CP000061	CP000061	CP000061	CP000061	CP000061	CP000061
CPA	australia grape yellow	Australia	16Sr XII	AM422018	AM422018	AM422018	AM422018	AM422018	AM422018	AM422018	AM422018	AM422018	AM422018	AM422018
CPM	apple proliferation	Germany	16Sr X	CU469464	CU469464	CU469464	CU469464	CU469464	CU469464	CU469464	CU469464	CU469464	CU469464	CU469464
SLY	strawberry lethal yellow	New Zealand	16Sr XII	CP002548	CP002548	CP002548	CP002548	CP002548	CP002548	CP002548	CP002548	CP002548	CP002548	CP002548
AWB	alfalfa witches'-broom		16Sr II	AY169322										
WX	western X		16Sr III	FJ376628										

(续表)

株系	引起的病害	采集地	级别	基因GenBank序列号									
				16S rDNA	rp	tuf	secA	secY	ipt	dnaK	fusA	gyrB	pyrG	rpoB
CLY	coconut lethal yellow		16SrIV	KF751388										
PPWB	pigeon pea witches'-broom		16SrIX	AB741637										
EY	elm yellow		16SrV	AF122911										
PWB	potato witches'-broom		16SrVI	DQ256089										
EWB	erigeron witches'-broom		16SrVII	AF411592										
LWB	loofah witches'-broom		16SrVIII	AF248956										
Stolbur	stolbur		16SrXII	AF248959										
BGWL	bermuda grass white leaf		16SrXIV	AF248961										
HWB	hibiscus witches'-broom		16SrXV	AF147708										
CWB-Y3	chinaberry witches'-broom	Hainan, China	16SrI	EF990733										
LY-TX7	lettuce yellow	Texas, U.S.A.	16SrI	KF573449										
MD-XT	mulberry dwarf	Shandong, China	16SrI	FJ844442										
PaWB-Korea	paulownia witches'-broom	Korea	16SrI	AB693131										
PeV-NA	periwinkle virescence	Italy	16SrI	HM590621										

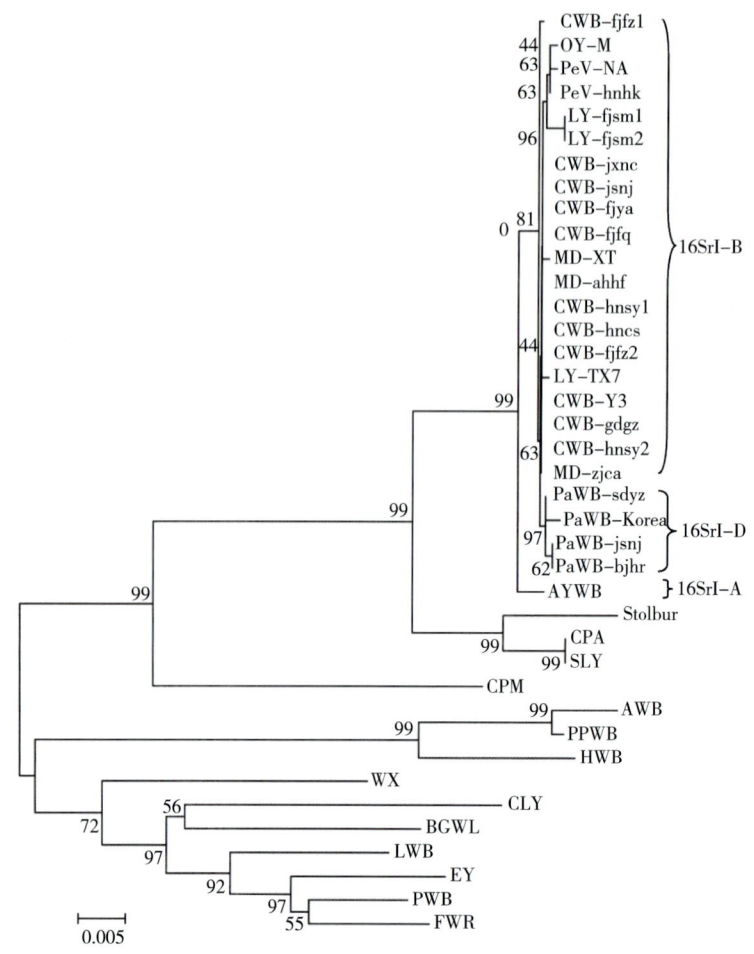

图3-23 基于16S rRNA基因序列构建的植原体NJ系统发育树（于少帅，2016）

注：枝长代表遗传距离，自展值（1 000个重复）在分枝处表示。

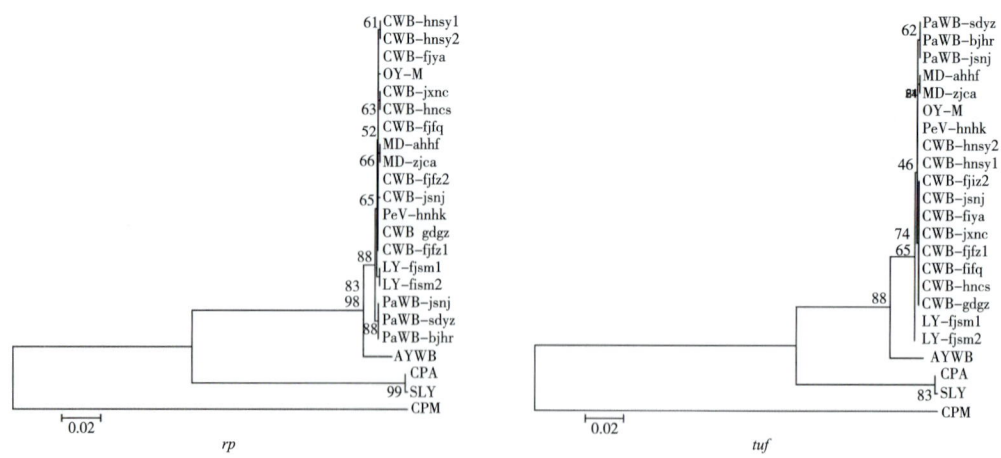

图3-24 基于植原体 *rp*、*tuf*、*secA*、*secY*、*ipt*、*dnaK*、*fusA*、*gyrB*、*pyrG*、*rpoB*基因序列构建的系统进化树（于少帅，2016）

注：枝长代表遗传距离，自展值（1 000个重复）在分枝处表示。

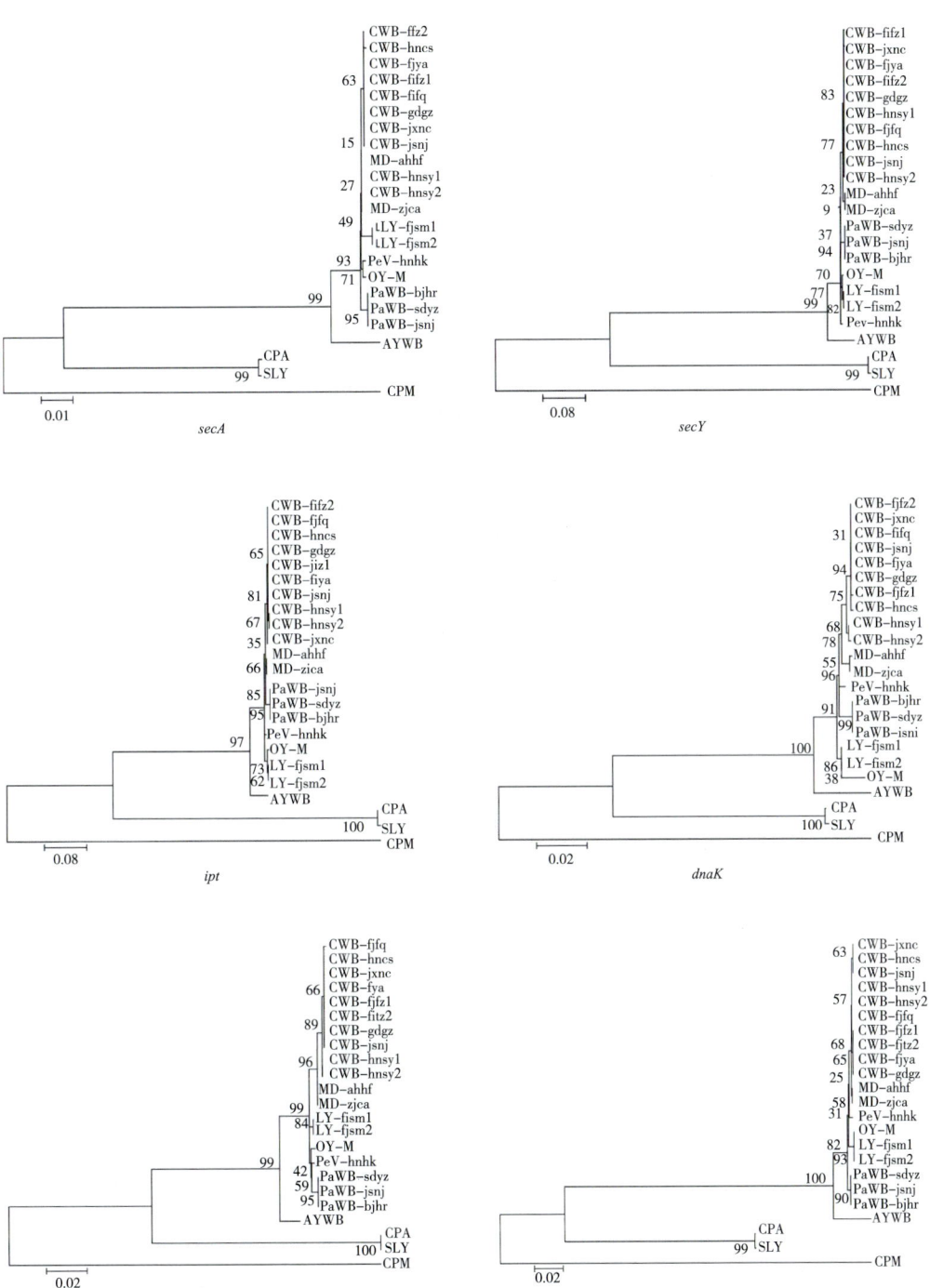

图3-24 基于植原体*rp*、*tuf*、*secA*、*secY*、*ipt*、*dnaK*、*fusA*、*gyrB*、*pyrG*、*rpoB*基因序列构建的系统进化树（于少帅，2016）（续）

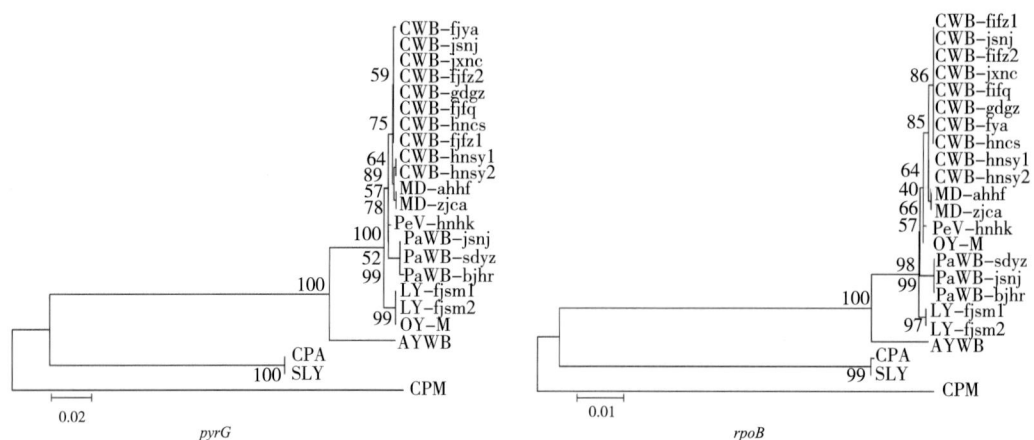

图3-24 基于植原体*rp*、*tuf*、*secA*、*secY*、*ipt*、*dnaK*、*fusA*、*gyrB*、*pyrG*、*rpoB*基因序列构建的系统进化树（于少帅，2016）（续）

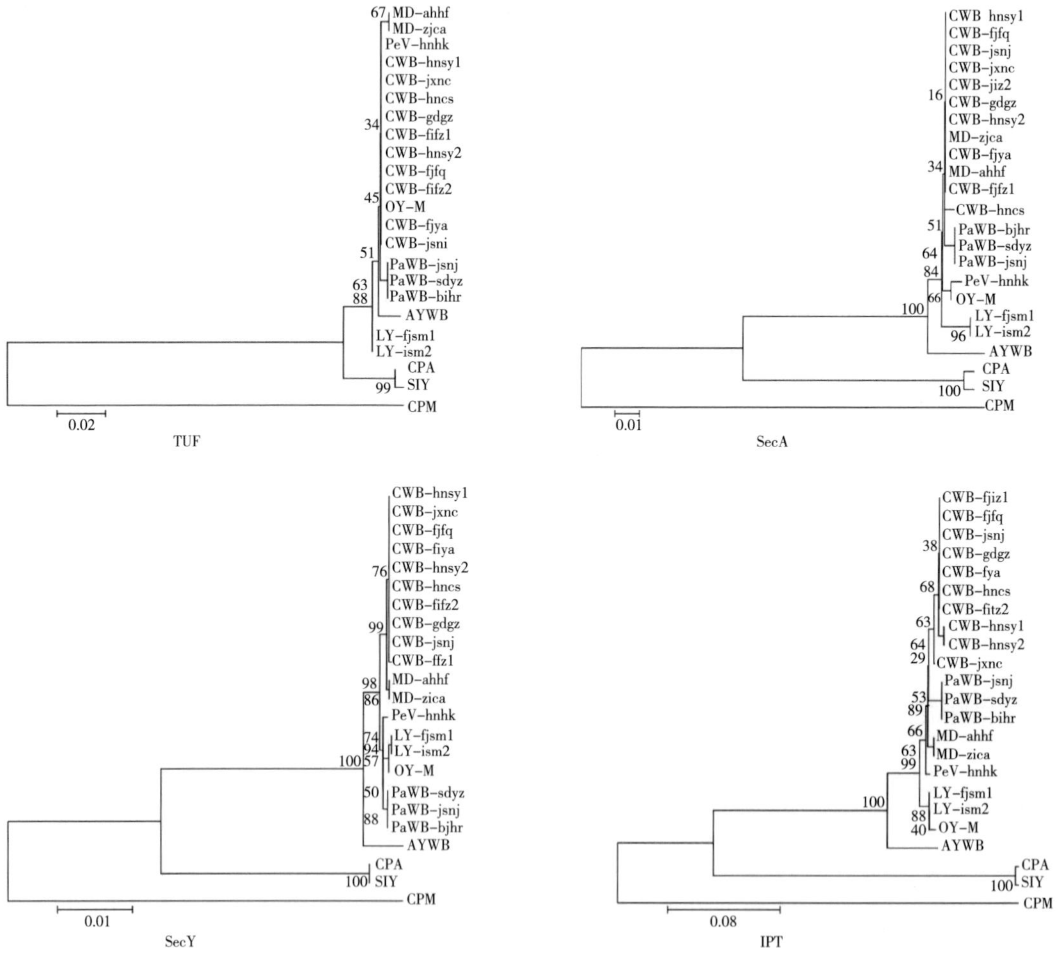

图3-25 基于植原体TUF、SecA、SecY、IPT、DnaK、FusA、GyrB、PyrG、RpoB蛋白序列构建系统发育树（于少帅，2016）

注：枝长代表遗传距离，自展值（1 000个重复）在分枝处表示。

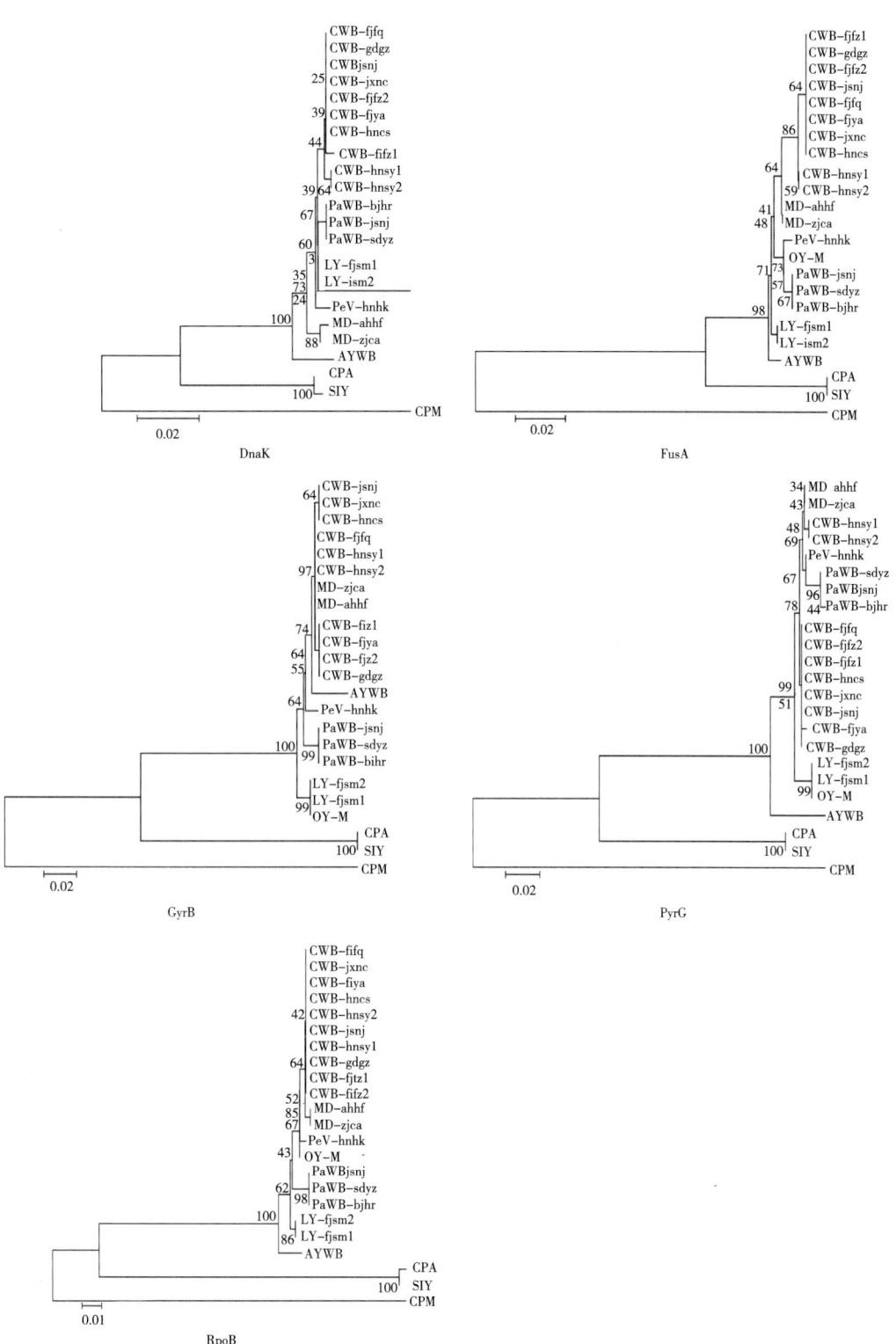

图3-25　基于植原体TUF、SecA、SecY、IPT、DnaK、FusA、GyrB、PyrG、RpoB蛋白序列构建系统发育树（于少帅，2016）（续）

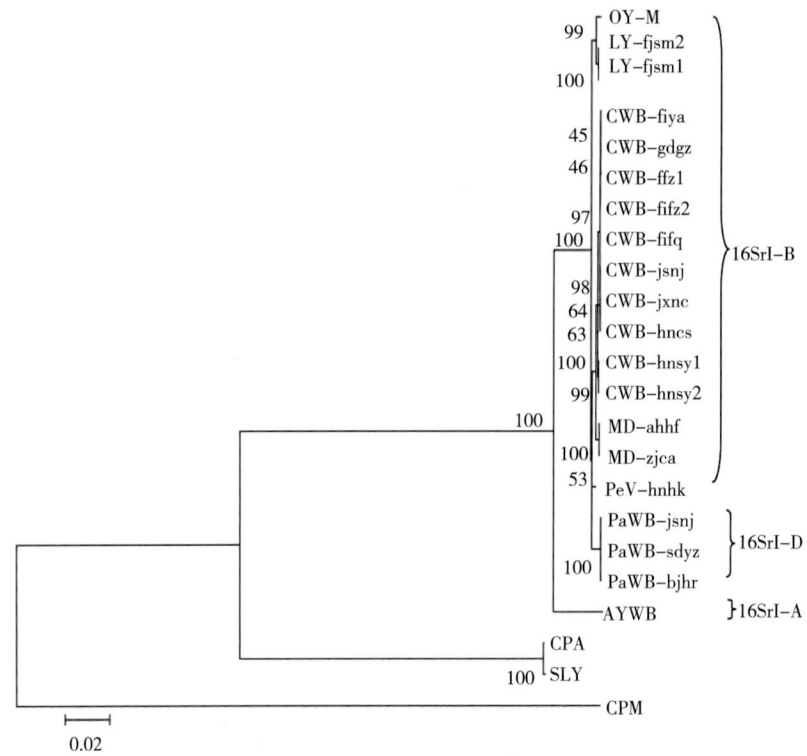

图3-26 基于16SrⅠ组和其他组的植原体株系 *rp*、*tuf*、*secA*、*secY*、*ipt*、*dnaK*、*fusA*、*gyrB*、*pyrG*、*rpoB* 基因序列的整合序列构建的系统进化树（于少帅，2016）

注：枝长代表遗传距离，自展值（1 000个重复）在分枝处表示。

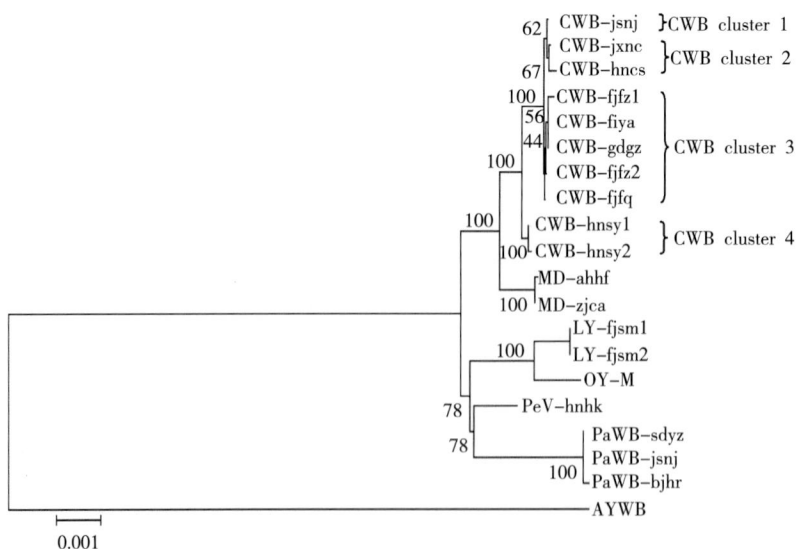

图3-27 基于16SrⅠ组植原体株系 *rp*、*tuf*、*secA*、*secY*、*ipt*、*dnaK*、*fusA*、*gyrB*、*pyrG*、*rpoB* 基因序列的整合序列构建的系统进化树（于少帅，2016）

注：枝长代表遗传距离，自展值（1 000个重复）在分枝处表示。

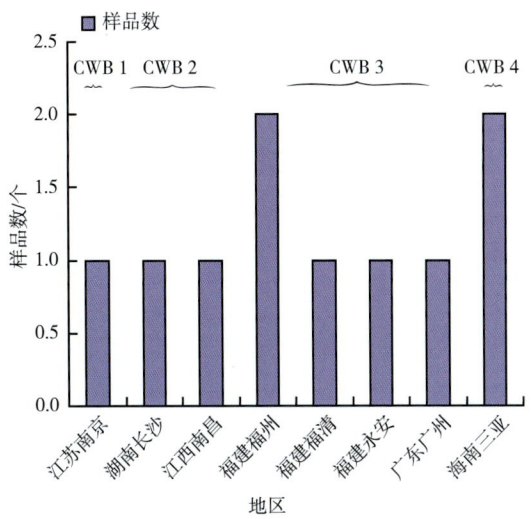

图3-28 我国4个苦楝丛枝植原体（CWB）分支中植原体株系地理分布（于少帅，2016）

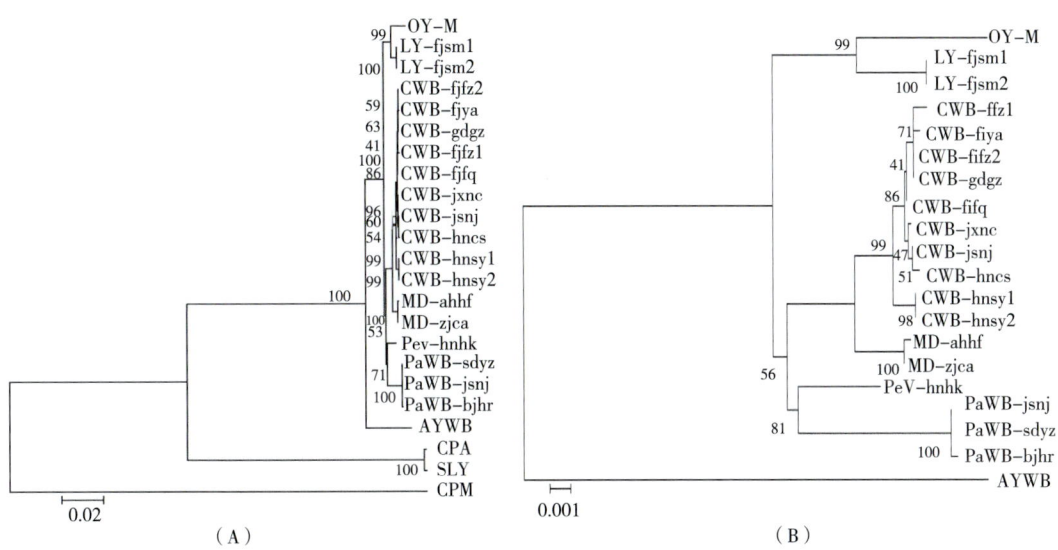

图3-29 基于植原体株系TUF、SecA、SecY、IPT、DnaK、FusA、GyrB、PyrG、RpoB蛋白序列的整合序列构建系统发育树（于少帅，2016）

注：基于推测的蛋白序列建立的进化树，（A）包括16SrⅠ组和其他组植原体株系；（B）只包含16SrⅠ组植原体株系。

十、核酸杂交检测技术

植原体核酸杂交技术是对植原体DNA片段进行扩增以制备特异性标记探针，通过核酸杂交反应检测植原体。1987年，Kirkpatrick等通过重组DNA技术克隆到西方桃X病植原体DNA，并在此基础上建立了点印迹杂交技术用于植原体检测，这是核酸杂交技术在植原体检测方面的应用。王宇等（2002）应用核酸点印记杂交法对枣疯病植原体进行了检测。核酸杂交技术不仅可对植原体进行检测，还可原位显示植原体在不同寄主、不同部位的分布情况。Webb等（1999）将标记的寡核苷酸作为杂交探针，基于核酸杂交技术对叶蝉体内的植原体进行了原位杂交，以此揭示不同时间段植原体在媒介昆虫中的位置变化。

十一、免疫检测技术

植原体免疫检测是基于抗原抗体反应进行的，抗原表面抗原决定簇可以诱导免疫动物淋巴细胞产生与抗原结合的特异性抗体，因此通过免疫反应可对植原体进行检测。1974年，Sinha等从感病的翠菊黄化组织中提纯植原体并制备了抗体，首次开展了植原体血清学研究。由于植原体尚难从植物中分离培养，植原体免疫抗原的制备难度较大。早期通过组织匀浆、差速离心等方式从感病植物中提纯植原体，但其中难免会含有植物成分，以此方法制备的抗体，在检测应用中易出现较高的假阳性。单克隆抗体制备技术为解决抗原纯化和高质量抗体制备提供了新的思路。1985年，Lin等通过单克隆抗体制备技术成功制备了AYWB植原体的单克隆抗体。由于单克隆抗体具有特异性强、灵敏度高等特点，所制备的AYWB植原体单克隆抗体具有高度专化性，只能与制备该抗体的植原体株系进行免疫反应，而不适用于其他植原体株系的检测。随后10余种植原体的单克隆抗体和多克隆抗体被制备（Garnier et al.，1990；Seddas et al.，1996；Shahriyari et al.，2011；Hodgetts et al.，2014）。通过植原体基因原核表达，以原核表达的蛋白制备抗原是植原体血清学研究上的又一项成功的尝试。2004年Wei等报道OY-secA兔抗血清与属于4个16Sr组的莴苣黄化、桑树萎缩、枣疯病等8个植原体株系有免疫反应。利用此方法，洋葱黄化和苹果丛生植原体免疫膜蛋白（Imp）抗血清，泡桐丛枝植原体的抗原膜蛋白、蛋白延伸因子、质粒ORF4以及葡萄黄化三磷酸甘油醛脱氢酶（GapA）等抗血清相继制备完成（Kakizawa et al.，2004；Lin et al.，2009；Wang et al.，2010；牟海青等，2011）。基于免疫学方法建立的植原体检测方法，操作简单，成本低，但一些抗体制备困难，成本高，易与寄主产生非特异性反应。

参考文献

曹学仁，车海彦，杨毅，等，2014. 基于Maxent的椰子致死性黄化植原体在中国的适生性分析[J]. 热带作物学报，35（11）：2260-2265.

车海彦，2010. 海南省植原体病害多样性调查及槟榔黄化病植原体的分子检测技术研究[D]. 杨凌：西北农林科技大学.

车海彦，罗大全. 一种检测槟榔黄化病植原体病原的方法及其专用试剂盒：ZL 2009 1

0077044. 3[P]. 2011-10-05.

车海彦, 吴翠婷, 符瑞益, 等, 2010. 海南槟榔黄化病病原物的分子鉴定[J]. 热带作物学报, 31（1）: 83-87.

金开璇, 蔡希灼, 阿·丝·诺耒斯, 1982. 苦楝丛枝病类细菌（BLO）、类菌原体（MLO）的电镜观察[J]. 林业科学, 18（4）: 422-424.

胡佳续, 宋传生, 林彩丽, 等, 2013. 4种植物病害植原体病原质粒全序列测定及分子特征[J]. 林业科学, 49（4）: 90-97.

蒯元璋, 2012. 桑树病原原核生物及其病害的研究进展（Ⅱ）[J]. 蚕业科学, 38（5）: 898-913.

赖帆, 李永, 徐启聪, 等, 2008. 植原体的最新分类研究动态[J]. 微生物学通报, 35（2）: 291-295.

李章宝, 1992. 我国桑树萎缩病的研究概况[J]. 湖南农业科学（6）: 41-43.

林积秀, 严叔平, 陈红运, 等, 2013. 福建永安莴苣上一种新病害的研究简报[J]. 中国植保导刊, 33（11）: 33-34.

卢全有, 2010. 桑树黄化型萎缩病植原体延伸因子基因的克隆及序列分析[J]. 植物保护, 36（5）: 43-46.

牟海青, 周涛, 赵文军, 等, 2011. 泡桐丛枝植原体抗原膜蛋白抗血清的制备及应用[J]. 植物病理学报, 41（2）: 161-170.

任争光, 王合, 林彩丽, 等, 2015. 实时荧光定量PCR（SYBR Green Ⅰ）检测不同抗枣疯病枣树品种嫁接接穗中的植原体浓度[J]. 植物病理学报, 45（5）: 520-529.

田国忠, 黄钦才, 1994. 感染MLO泡桐组培苗代谢变化与致病机理的关系[J]. 中国科学（5）: 484-490.

王纯利, 袁自清, 吕巡贤, 等, 1992. 新疆柳树丛枝病类菌原体（MLO）的电镜观察[J]. 新疆农业大学学报（4）: 17-19.

王洁, 田国忠, 徐启聪, 等, 2010. 泡桐丛枝病树周围几种植物上植原体的分子检测[J]. 中国农业科学, 43（2）: 304-312.

王宇, 2002. 核酸点印迹杂交法检测枣疯病植原体[J]. 河北林业科技（3）: 1-3.

王圣洁, 2017. 我国重要林木植原体分子检测新技术的研发和遗传多样性研究[D]. 北京: 中国林业科学研究院.

王圣洁, 林彩丽, 严东辉, 等, 2017a. 寡核苷酸管芯片技术检测和鉴别我国不同组植原体[J]. 林业科学研究, 30（1）: 99-110.

王圣洁, 王胜坤, 林彩丽, 等, 2017b. 以tuf基因为靶标的5种16SrⅠ组植原体环介导恒温扩增技术[J]. 林业科学, 53（8）: 54-63.

于少帅, 2016. 植原体tuf基因启动子分子特征和枣树抗植原体物质研究[D]. 北京: 中国林业科学研究院.

于少帅，林彩丽，王圣洁，等，2018. 植原体 *tuf* 基因与其上游部分基因结构和相关基因启动子保守区域特征及活性分析[J]. 生物多样性，26（7）：738-748.

于少帅，宋薇薇，覃伟权，2021. 海南槟榔黄化植原体分子检测及其系统发育关系研究[J]. 热带作物学报，42（11）：3066-3072.

朱辉，覃伟权，吴多扬，等，2010. 椰子致死性黄化植原体传入中国的风险性分析[J]. 江西农业学报，22（11）：84-87.

ARNAUD G，MALEMBIC-MAHER S，SALAR P，et al.，2007. Multilocus sequence typing confirms the close genetic interrelatedness of three distinct Flavescence Dorée phytoplasma strain clusters and group 16SrⅤ phytoplasmas infecting grapevine and alder in Europe[J]. Applied and Environmental Microbiology，73（10）：4001-4010.

BABU M，THANGESWARI S，JOSEPHRAJKUMAR A，et al.，2021. First report on the association of 'Candidatus Phytoplasma asteris' with lethal wilt disease of coconut（*Cocos nucifera* L.）in India[J]. Journal of General Plant Pathology，87：16-23.

BARBA M，HADIDI A，RAO G，et al.，2008. DNA microarrays：technology，applications and potential applications for the detection of plant viruses and virus-like pathogens[J]. Techniques in diagnosis of plant viruses：227-247.

BERTACCINI A，DUDUK B，PALTRINIERI S，et al.，2014. Phytoplasmas and phytoplasma diseases：a severe threat to agriculture[J]. American Journal of Plant Sciences，5：1763-1788.

BISHOP C J，AANENSEN D M，JORDAN G E，et al.，2009. Assigning strains to bacterial species via the internet[J]. BMC Biology，7（1）：3.

CAMPAS M，KATAKIS I，2004. DNA biochip arraying，detection and amplification strategies[J]. TrAC Trends in Analytical Chemistry，23：49-62.

CHRISTENSEN N M，NICOLAISEN M，HANSEN M，et al.，2004. Distribution of phytoplasmas in infected plants as revealed by real-time PCR and bioimaging[J]. Molecular Plant-microbe Interactions：MPMI，17（11）：1175-1184.

DANET J L，BONNET P，JARAUSCH W，et al.，2007. *Imp* and *secY*，two new markers for MLST（multilocus sequence typing）in the 16SrⅩ phytoplasma taxonomic group[J]. Bulletin of Insectology，60（60）：339-340.

DAY E，DEAR P H，MCCAUGHAN F，2013. Digital PCR strategies in the development and analysis of molecular biomarkers for personalized medicine[J]. Methods，59（1）：101-107.

DENG S，HIRUKI C，1991. Genetic relatedness between two nonculturable mycoplasma like organisms revealed by nucleic acid hybridization and polymerase chain reaction[J]. Phytopathology, 81（12）: 1475-1479.

DICKINSON M, HODGETTS J, 2013. Phytoplasma: methods and protocols[D]. Totowa: Humana Press.

DOI Y, TERANAKA M, YORA K, et al., 1967. Mycoplasma or PLT group like microorganisms found in the phloem elements of pants infected with mulberry dwarf, potato witches'-broom, aster yellows or pauwlonia witches'-broom[J]. Annals of the Phytopathological Society of Japan, 33: 259-266.

HODGETTS J, BOONHAM N, MUMFORD R, et al., 2008. Phytoplasma phylogenetics based on analysis of secA and 23S rRNA gene sequences for improved resolution of candidate species of 'Candidatus Phytoplasma' [J]. International Journal of Systematic and Evolutionary Microbiology, 58（8）: 1826-1837.

HODGETTS J, JOHNSON G, PERKINS K, et al., 2014. The development of monoclonal antibodies to the secA protein of Cape St. Paul wilt disease phytoplasma and their evaluation as a diagnostic tool[J]. Molecular Biotechnology, 56（9）: 803-813.

GARNIER M, MARTINGROS G, ISKRA M L, et al., 1990. Monoclonal anitbodies against the MLOs associated with tomato stolbur and clover phyllody[J]. Zentralblatt Fur Bakteriologie Supplement, 28（9）: 82-86.

GEVERS D, COHAN F M, LAWRENCE J G, et al., 2005. Opinion: re-evaluating prokaryotic species[J]. Nature Reviews Microbiology, 3（9）: 733-739.

GUNDERSEN D E, LEE I M, 1996. Ultrasensitive detection of phytoplasmas by nested-PCR assays using two universal primer pairs[J]. Phytopathologia Mediterranea, 35: 144-151.

GURR G M, JOHNSON A C, ASH G J, et al., 2016. Coconut lethal yellowing diseases: a phytoplasma threat to palms of global economic and social significance[J]. Frontiers in Plant Science, 7: 1521.

KAKIZAWA S, OSHIMA K, NISHIGAWA H, et al., 2004. Secretion of immunodominant membrane protein from onion yellows phytoplasma through the Sec protein-translocation system in Escherichia coli[J]. Microbiology, 150（1）: 135-142.

KIRKPATRICK B C, STENGER D C, MORRIS T J, et al., 1987. Cloning and detection of DNA from a nonculturable plant pathogenic mycoplasma-like organism[J]. Science, 238（4824）: 197.

LEE I M, GUNDERSEN-RINDAL D E, DAVIS R E, et al., 2004. 'Candidatus Phytoplasma asteris', a novel phytoplasma taxon associated with aster yellows and related diseases[J]. International Journal of Systematic and Evolutionary Microbiology, 54（4）:

1037-1048.

LEE I M, HAMMOND R W, DAVIS R E, et al., 1993. Universal amplification and analysis of pathogen 16S rDNA for classification and identification of mycoplasmalike organisms[J]. Phytopathology, 83: 834-842.

LEE I M, MARTINI M, BOTTNER K D, et al., 2003. Ecological implications from a molecular analysis of phytoplasmas involved in an aster yellows epidemic in various crops in Texas[J]. Phytopathology, 93 (11): 1368-1377.

LI Y, PIAO C G, TIAN G Z, et al., 2014. Multilocus sequences confirm the close genetic relationship of four phytoplasmas of peanut witches'-broom group 16SrⅡ-A[J]. Journal of Basic Microbiology, 54 (8): 818-827.

LIN C P, AN C T, 1985. Monoclonal antibodies against the aster yellows agent[J]. Science, 227 (4691): 1233-1235.

LIN J X, MOU H Q, LIU J M, et al., 2014. First report of lettuce chlorotic leaf rot disease caused by phytoplasma in China[J]. Plant Disease, 98 (10): 1425.

LIN C, ZHOU T, LI H, et al., 2009. Molecular characterisation of two plasmids from paulownia witches'-broom phytoplasma and detection of a plasmid-encoded protein in infected plants[J]. European Journal of Plant Pathology, 123 (3): 321.

MAIDEN M C J, BYGRAVES J A, FEIL E, et al., 1998. Multilocus sequence typing: a portable approach to the identification of clones within populations of pathogenic microorganisms[J]. Proceedings of the National Academy of Sciences of the United States of America, 95 (6): 3140-3145.

MULLIS K B, FALOONA F A, 1987. Specific synthesis of DNA in vitro via a polymerase-catalyzed chain reaction[J]. Methods in Enzymology, 155: 335-350.

MUSIC M C, PUSIC P, FABRE A, et al., 2011. Variability of stolbur phytoplasma strains infecting Croatian grapevine by multilocus sequence typing[J]. Bulletin of Insectology, 64: S39-S40.

MYLES S, MAHANIL S, HARRIMAN J, et al., 2015. Genetic mapping in grapevine using SNP microarray intensity values[J]. Molecular Breeding, 35: 1-12.

NAIR S, MANIMEKALAI R, RAJ P G, et al., 2016. Loop mediated isothermal amplification (LAMP) assay for detection of coconut root wilt disease and arecanut yellow leaf disease phytoplasma[J]. World Journal of Microbiology and Biotechnology, 32 (7): 1-7.

NICOLAISEN M, BERTACCINI A, 2007. An oligonucleotide microarray-based assay for identification of phytoplasma 16S ribosomal groups[J]. Plant Pathology, 56: 332-336.

NOTOMI T, OKAYAMA H, MASUBUCHI H, et al., 2000. Loop-mediated isothermal amplification of DNA[J]. Nucleic Acids Research, 28（12）: e63.

OBURA E, MASIGA D, WACHIRA F, et al., 2011. Detection of phytoplasma by loop-mediated isothermal amplification of DNA（LAMP）[J]. Journal of Microbiological Methods, 84（2）: 312-316.

SAUX M F, 2013. What new technologies bring to challenge of bacterial classification: unraveling evolution and ecology of plant pathogenic bacteria[J]. Acta Phytopathologica Sinica, 43: 538.

SCHNEIDER B, GIBB K S, SEEMÜLLER E, 1997. Sequence and RFLP analysis of the elongation factor Tu gene used in differentiation and classification of phytoplasmas[J]. Microbiology, 143（10）: 3381-3389.

SEDDAS A, MEIGNOZ R, DAIRE X, et al., 1996. Generation and characterization of monoclonal antibodies to flavescence dorée phytoplasma: serological relationships and differences in electroblot immunoassay profiles of flavescence dorée and elm yellows phytoplasmas[J]. European Journal of Plant Pathology, 102（8）: 757-764.

SHAHRIYARI F, SAFARNEJAD M R, SHAMSBAKHSH M, et al., 2011. Generation of a specific monoclonal recombinant antibody against 'Candidatus Phytoplasma aurantifolia' using phage display technology[J]. Bulletin of Insectology, 64: 75-76.

SINHA R C, 1974. Purification of mycoplasma-like organisms from China aster plants affected with clover phyllody[J]. Phytopathology, 64（8）: 1156.

SPRATT B G, 1999. Multilocus sequence typing: molecular typing of bacterial pathogens in an era of rapid DNA sequencing and the internet[J]. Current Opinion in Microbiology, 2（3）: 312-316.

SUGAWARA K, HIMENO M, KEIMA T, et al., 2012. Rapid and reliable detection of phytoplasma by loop-mediated isothermal amplification targeting a housekeeping gene[J]. Journal of General Plant Pathology, 78（6）: 389-397.

TIAN G Z, RAYCHAUDHURI S P, 1996. Paulownia witches'-broom disease in China: present status[M]//Raychaudhuri RS, Moromorosch K, eds. Forest Trees and Palms: Diseases and Control. New Delhi: Oxford & IBH Publishing Company, 227-251.

THOMAS S, BALASUNDARAN M, 1998. In situ detection of phytoplasma in spike disease affected sandal using DAPI stain[J]. Current Science, 74（11）: 989-993.

TORRES E, BERTOLINI E, CAMBRA M, et al., 2005. Real-time PCR for simultaneous and quantitative detection of quarantine phytoplasmas from apple proliferation（16SrⅩ）group[J]. Molecular and Cellular Probes, 19（5）: 334-340.

WANG J, ZHU X P, GAO R, et al., 2010. Genetic and serological analyses of elongation factor EF-Tu of paulownia witches'-broom phytoplasma (16SrⅠ-D) [J]. Plant Pathology, 59(5): 972-981.

WEBB D R, BONFIGLIOLI R G, CARRARO L, et al., 1999. Oligonucleotides as hybridization probes to localize phytoplasmas in host plants and insect vectors[J]. Phytopathology, 89(10): 894.

WEI W, KAKIZAWA S, JUNG H Y, et al., 2004. An antibody against the SecA membrane protein of one phytoplasma reacts with those of phylogenetically different phytoplasmas[J]. Phytopathology, 94(7): 683-686.

YIP T T, KWONG D L, NGAN R K, et al., 2015. Differential transcript expression in nasopharyngeal carcinoma by cDNA microarray analysis[J]. Cancer Research, 75: 3903-3903.

YU S S, CHE H Y, WANG S J, et al., 2020. Rapid and efficient detection of 16SrⅠ group areca palm yellow leaf phytoplasma in China by loop-mediated isothermal amplification[J]. The Plant Pathology Journal, 36(5): 459-467.

YU S S, LI Y, REN Z G, et al., 2017. Multilocus sequence analysis for revealing finer genetic variation and phylogenetic interrelatedness of phytoplasma strains in 16SrⅠ group in China[J]. Scientia Silvae Sinicae, 53(3): 105-118.

YU S S, PAN Y W, ZHU H, et al., 2023. Universal, Rapid and visual detection methods for phytoplasmas associated with coconut lethal yellow-type diseases targeting 16S rRNA gene sequences[J]. Plant Disease, 107(2): 276-280.

YU S S, ZHANG X C, SONG W W, et al., 2022a. Accurate and sensitive detection of areca palm yellow leaf phytoplasma in China by droplet digital PCR targeting *tuf* gene sequence[J]. Annual of Applied Biology, 181: 152-159.

YUAN X L, MORANO L, BROMLEY R, et al., 2010. Multilucus sequence typing of *Xylella fastidiosa* causing Pierce's disease and oleander leaf scorch in the United States[J]. Phytopathology, 100(6): 601-611.

第四章

植原体致病机理

植原体主要通过分泌效应因子蛋白与植物寄主的靶标蛋白互作，从而导致寄主植物病害的发生，这些效应因子蛋白可以干扰寄主植物细胞内的物质转运、基因表达和防御反应等过程（Oshima et al.，2004；Bai et al.，2006；Tran-Nguyen et al.，2008；Kube et al.，2008；Wang et al.，2018a；Chen et al.，2014；Orlovskis et al.，2017；Music et al.，2019；Chen et al.，2022；Hoshi et al.，2009；Huang et al.，2021；李继东等，2019；于少帅等，2016）。致病效应因子需要通过细胞膜进入寄主细胞发挥作用。植原体致病因子分泌主要通过Sec分泌系统完成，植原体编码Sec蛋白转运系统核心功能蛋白，如secA、secY等，能够识别并切割位于蛋白质N端保守的信号肽序列，从而将致病因子分泌并整合到植物寄主的细胞质中（Bai et al.，2006；Kakizawa et al.，2004）。通过SOSUI、SignalP或PSORT等软件分析预测蛋白的信号肽序列，TMHMM软件分析预测跨膜结构域，含信号肽序列但不含跨膜结构域的蛋白即为分泌蛋白，也是候选的致病效应因子。

一、植原体效应子

植原体全基因组测序目前已完成的有洋葱黄化植原体（Onion yellow，OY-M）、翠菊黄化丛枝植原体（Aster yellow witches'-broom，AYWB）、澳大利亚葡萄黄化植原体（Australian grapevine yellows，AUSGY）、草莓致死黄化植原体（Strawberry lethal yellow，SLY）、苹果簇生植原体（Apple proliferation，AP）和枣疯病植原体（Jujube witches'-broom，JWB）等（Oshima et al.，2004；Bai et al.，2006；Tran-Nguyen et al.，2008；Kube et al.，2008；Andersen et al.，2013；Wang et al.，2018a）；此外，小麦蓝矮植原体等19个植原体完成了基因组草图测序。基于植原体基因组序列和生物信息学软件预测分析得知，OY-M植原体编码30个分泌蛋白，AYWB植原体编码56个分泌蛋白，AUSGY植原体编码41个分泌蛋白，AP植原体编码13个分泌蛋白，JWB植原体编码28个分泌蛋白，小麦蓝矮植原体编码37个分泌蛋白，玉米丛枝矮化植原体编码36个分泌蛋白，

ZjTCP7的表达诱导枣苗丛枝表型产生，如图4-1所示。其机理可能是低含量的ZjTCP7蛋白降低了独脚金内酯合成相关基因的转录水平导致的（Chen et al.，2022）。

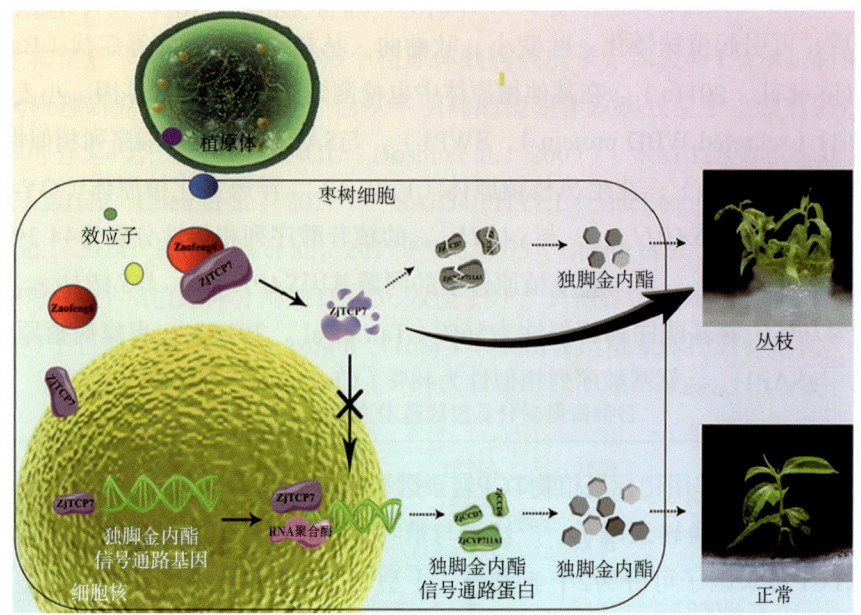

图4-1　枣疯病植原体Zaofeng6效应子致病机理示意图（Chen et al.，2022）

2. SAP54类效应子

SAP54（secreted AYWB protein 54）是AYWB植原体分泌的效应蛋白，是植物花变叶的诱导蛋白，编码这类蛋白的基因被命名为叶变基因（phyllody inducing gene，phyllogen）（MacLean et al.，2011）。叶变基因编码的蛋白含有125个氨基酸（Maejima et al.，2014）。在洋葱黄化植原体中鉴定到的该类基因为$PHYL1_{OY}$，其编码蛋白的氨基酸序列与SAP54相似性为88%（Maejima et al.，2014）；在花生丛枝植原体中鉴定到的该类基因为$PHYL1_{PnWB}$，与SAP54氨基酸序列相似性为61.3%，与$PHYL1_{OY}$氨基酸序列相似性为56%（Kitazawa et al.，2017）。

SAP54效应子与植物的互作靶标为MADS转录因子，MADS转录因子对调控植物花器官的形成具有重要作用（MacLean et al.，2014；Maejima et al.，2014；Kitazawa et al.，2017）。植原体SAP54效应子与寄主植物的A类和E类MADS转录因子互作，通过泛素-蛋白酶体途径降解这些转录因子，使寄主植物花器官不能正常生长，导致花变叶症状的产生（MacLean et al.，2014；Maejima et al.，2014；Kitazawa et al.，2017）。酵母双杂交实验表明，phyllogens基因编码的蛋白可以与拟南芥、烟草、长春花等植物的MADS转录因子互作，导致花发育畸形、花瓣变绿、花器变叶等症状（Maejima et al.，2014；Kitazawa et al.，2017）。phyllogens基因编码蛋白也能和日本柳杉（*Cryptomeria japonica*）、欧洲云杉（*Picea abies*）等裸子植物和粗梗水蕨（*Ceratopteris pterioides*）等蕨类植物的MADS

转录因子互作（Kitazawa et al.，2017）。phyllogens基因在拟南芥中表达能吸引媒介昆虫的定殖，有利于植原体的传播（MacLean et al.，2014；Maejima et al.，2014；Kitazawa et al.，2017）。

3. SAP05效应子

SAP05效应子是AYWB植原体分泌的一类蛋白，控制着植物的多个发育过程（Bai et al.，2008；Huang et al.，2021）。植原体的这类效应因子可以延长植物寄主的寿命，同时诱导植原体和其载体定殖的组织器官如叶和芽等产生丛枝状增殖和不育（Huang et al.，2021）。SAP05效应因子可以依赖于底物泛素化劫持植物泛素受体RPN10，通过非泛素依赖性蛋白降解途径调节并降解植物转录因子SPL和GATA，从而诱导植物产生丛枝症状（Huang et al.，2021）。泛素受体RPN10在真核生物中高度保守，但SAP05只结合植物的泛素受体RPN10，不结合昆虫载体的泛素受体RPN10，如图4-2所示（Huang et al.，2021）。基于昆虫的基因工程分析表明植物泛素受体RPN10中的2个氨基酸被取代可以产生对植原体SAP05效应因子活性具有抗性的功能变体（Huang et al.，2021）。

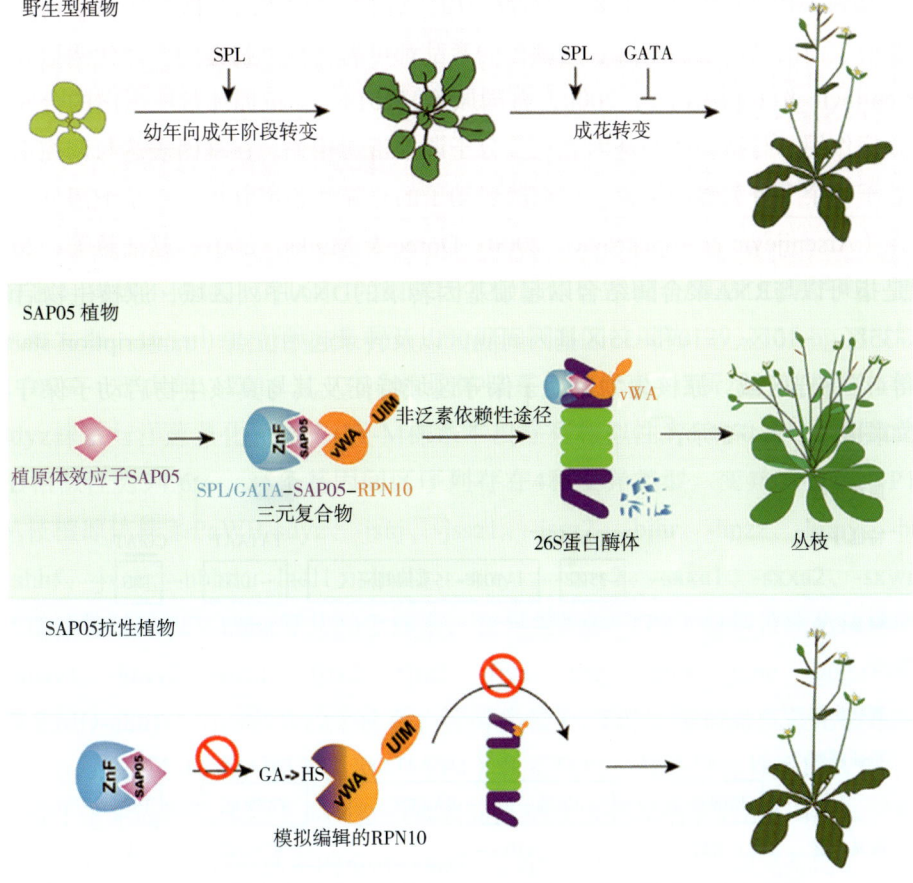

图4-2 植原体效应子SAP05致病机理模式图（Huang et al.，2021）

表4-3 不同植原体株系信息（于少帅，2016）

序号	植原体名称	组别	采样地和株系编号	株系数量
1	泡桐丛枝植原体	16SrⅠ	山东兖州（PaWB-sdyz），江苏南京（-jsnj）、苏州（-jssz1、-jssz2），北京怀柔（-bjhr），河南郑州（-hnzz）、濮阳（-hnpy）、洛阳（-hnly），湖南长沙（-hncs），安徽合肥（-ahhf），江西南昌（-jxnc），福建福州（-fjfz），河北保定（-hbbd），辽宁大连（-lndl1、-lndl2），山西太原（-sxty1、-sxty2），陕西西安（-sxxa1、-sxxa2）、渭南（-sxwn）	20
2	苦楝丛枝植原体	16SrⅠ	海南三亚（CWB-hnsy1、-hnsy2），福建福州（-fjfz1、-fjfz2、-fjfz3）、永安（-fjya），江苏南京（-jsnj），湖南长沙（-hncs），江西南昌（-jxnc），广东广州（-gdgz）	10
3	莴苣黄化植原体	16SrⅠ	福建永安（LY-fjya1、-fjya2）	2
4	桑萎缩植原体	16SrⅠ	安徽合肥（MD-ahhf），浙江淳安（-zjca）	2
5	长春花绿变植原体	16SrⅠ	海南海口（PeV-hnhk）	1

```
TP1        ACAACCTTACACTAAAAAACATTTTTATTAACAAAATAGT    40
TP2        ACAACCTTACACTAAAAAACATTTTTATTAACAAAATAGT    40
TP3        ACAACCTTACACTAAAAAACATTTTTATTAACAAAATATT    40
TP4        ACAACCTTAAACTAAAAAACATTTTTATTAACAAAATAGT    40
Consensus  acaacctta actaaaaaacattttt attaacaaaat t

TP1        TGTGAAAAGCCAAAAATAAGTTAAAATTTAAATGGTAAAT    80
TP2        TGTGAAAAGCCAAAAATAAGTTAAAATTTAAATGGTAAAT    80
TP3        TGTAAAAAGCCAAAAATAAGTTAAAATTTAAATGGTAAAT    80
TP4        TGTGAAAAGCCAAAAATAAGTTAAAATTTAAATGGTAAAT    80
Consensus  tgt aaaagccaaaaataagttaaaatttaaatggtaaat

TP1        ATATTTATTAAAACAAACCTTAAAACATAATAAAAGGAGG    120
TP2        ATATTT TTAAAACAAACCTTAAAACATAATAAAAGGAGG    119
TP3        ATATTTATTAAAACAAACCTTAAAACATAATAAAAGGAGG    120
TP4        ATATTTATTAAAACAAACCTTAAAACATAATAAAAGGAGG    120
Consensus  atattt ttaaaacaaaccttaaaacataataaaaggagg

TP1        CCTTTGAAAA                                   130
TP2        CCTTTGAAAA                                   129
TP3        CCTTTGAAAA                                   130
TP4        CCTTTGAAAA                                   130
Consensus  cctttgaaaa
```

图4-4 16SrⅠ组植原体*fusA*和*tuf*基因间区序列变异类型分析（于少帅，2016）

注：不同植原体株系*fusA*和*tuf*基因间区序列变异位点用竖框标出；序列中的"."表示碱基缺失。

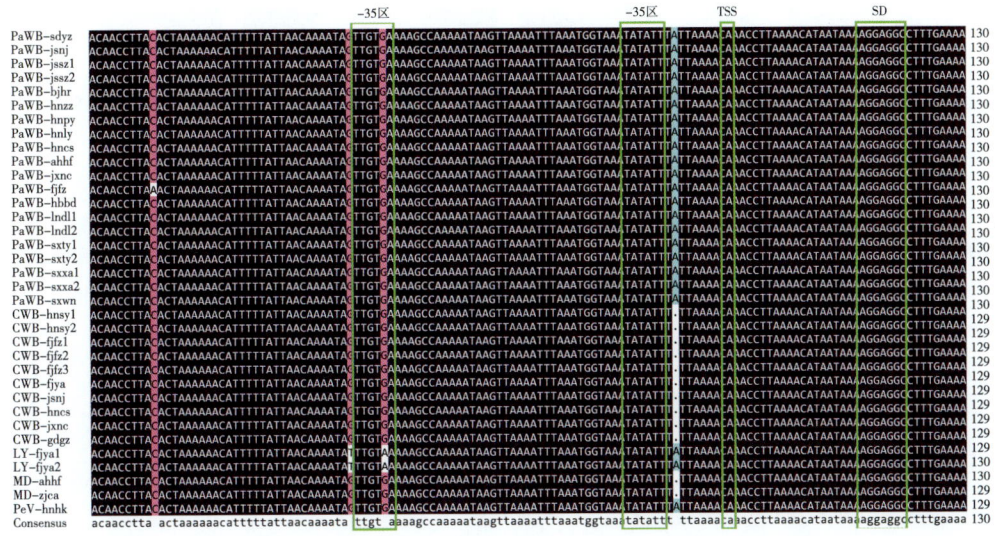

图4-5　16SrⅠ组株系fusA—tuf基因间区多重序列比对及启动子结构分析（于少帅，2016）

2. 启动子功能活性

为确定fusA—tuf基因间区序列启动子活性及序列变异对启动活性的影响，在16SrⅠ组植原体株系fusA和tuf基因间区4种变异类型中，每种变异类型选取1个代表株系构建启动子表达体系，用于启动子活性验证（于少帅，2016）。4种变异类型的代表株系分别是TP1为PaWB-sdyz，TP2为CWB-hnsy1，TP3为LY-fjya1，TP4为PaWB-fjfz；以泡桐丛枝枝植原体山东兖州株系PaWB-sdyz构建体系，将泡桐丛枝枝植原体山东兖州株系PaWB-sdyz tuf上游130 bp序列与启动子探针载体pSUPV4连接构建融合表达体系，在含30 mg/L和50 mg/L卡那霉素（Kan）的LB平板上培养24 h后，均有明显的转化子产生，挑取单菌落测序发现转化子中含有目的片段，因而判断该区域具有启动子结构和活性。将其他3种变异类型的代表性植原体株系的fusA和tuf基因间区与启动子探针载体pSUPV4连接构建融合表达体系，在含30 mg/L和50 mg/L Kan的LB平板上培养24 h后，观察表明，CWB-hnsy1、LY-fjya1和PaWB-fjfz LB平板上均有明显的转化子产生，挑取单菌落测序发现转化子中含有目的片段，因而判断3株植原体株系的该区域具有启动子结构和活性，如图4-6所示。

3. 植原体启动子保守区域特征

已知原核生物操纵子的基因结构和其调控序列的变化都会影响操纵子的表达。研究发现大肠杆菌rps12、rps7、fusA、tuf共同组成了一个str操纵子，rps12前和tuf后分别存在启动子和转录终止结构（Post et al.，1980）。大肠杆菌中rpS10基因在str操纵子下游15 kb处，与核糖体蛋白编码基因rpl3、rpl4等组成S10操纵子，在S10操纵子3'端发现转录终止结构（Post et al.，1980；Zurawski et al.，1985）。Sanangelantoni等（1993）研究发现，甲烷叶菌Methanococcus、Vannielii和Cyanelles等细菌的核糖体蛋白编码基因rpS10与tuf基因连在一起，与str操纵子一起转录。spc操纵子和S10操纵子在大肠杆菌

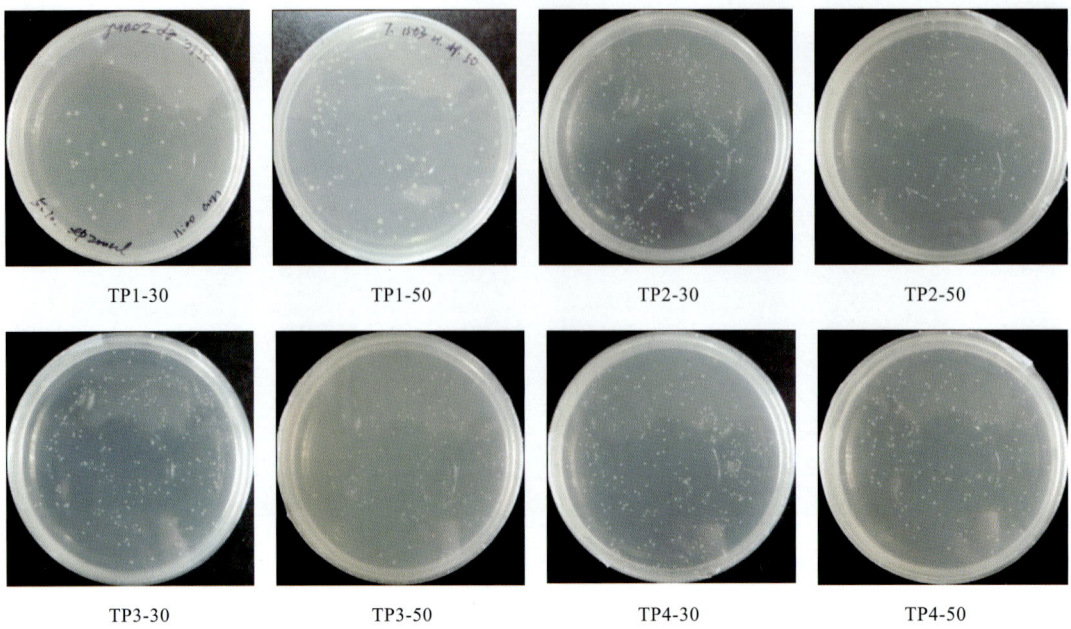

| TP1-30 | TP1-50 | TP2-30 | TP2-50 |
| TP3-30 | TP3-50 | TP4-30 | TP4-50 |

图4-6 融合表达体系在含Kan的LB培养基上的表达状况（于少帅，2016）

注：图片中的编号展示不同变异类型代表株系 $fusA$—tuf 基因间区的融合表达体系在不同Kan浓度LB平板上的生长情况。TP1—TP4代表4种变异类型，30、50代表LB平板上的Kan浓度（mg/L）。

基因组中均有发现（Post et al.，1980；Zurawski et al.，1985）。Miyata等（2002a）研究发现，植原体 $S10$ 操纵子和 spc 操纵子之间没有找到与转录起始或终止相关的序列，在 $S10$ 操纵子上游邻接序列发现了可能的启动子结构，推测 $S10$ 操纵子和 spc 操纵子在植原体中是一个转录单位，构成 $S10$-spc 操纵子。因而推断洋葱黄化植原体OY的 str 操纵子的结构为 5′-$rps12$-$rps7$-$fusA$-tuf-3′，与大肠杆菌的 str 操纵子结构排布一致。与支原体相比，洋葱黄化植原体OY的 str 操纵子基因结构排布与芽孢杆菌更接近（Miyata et al.，2002b），由此可见细菌基因结构及其调控的多样性和复杂性。

对16SrⅠ组植原体株系 tuf 基因启动子结构和活性进行分析时发现，泡桐丛枝植原体等16SrⅠ组植原体株系 $fusA$—tuf 基因间区长129～130 bp，通过softberry软件等方法预测发现存在 $fusA$—tuf 基因间区具有完整的启动子结构，经pSUPV4启动子探针检测到了启动子活性（于少帅，2016）。由此可见，泡桐丛枝植原体等16SrⅠ组植原体株系中的 tuf 基因在植原体中可能可以单独表达。采用高保真TaKaRa LA Taq 通过长片段PCR扩增我国泡桐丛枝植原体山东兖州株系PaWB-sdyz、福建福州株系PaWB-fjfz和莴苣黄化植原体福建永安株系LY-fjya1基因 tuf 及其上游6个基因 $rplL$、$rpoB$、$rpoC$、$rps12$、$rps7$、$fusA$ 序列，结合全基因测序已完成的16SrⅠ组的洋葱黄化植原体（onion yellows phytoplasma，OY-M）（GenBank登录号AP006628）和翠菊黄化丛枝植原体（aster yellows witches'-broom phytoplasma，AYWB）（CP000061），16SrⅫ组的澳大利亚葡萄黄化植原体（Candidatus

Phytoplasma australiense，PAa）（AM422018）和草莓致死黄化植原体（strawberry lethal yellows phytoplasma，SLY）（CP002548），还有16SrⅩ组的苹果簇生植原体（Candidatus Phytoplasma mali，AT）（CU469464）共5个株系的 *tuf* 及其上游6个基因 *rplL*、*rpoB*、*rpoC*、*rps12*、*rps7*、*fusA* 序列，以及非固醇甾原体PG-8A株系 *tuf* 及其上游基因序列进行比较分析，发现7个基因的结构顺序为5'-rplL-rpoB-rpoC-rps12-rps7-fusA-tuf-3'。每个基因对应的编码产物分别为：

  rplL：核蛋白L7/L12（ribosomal protein L7/L12）；

  rpoB：RNA聚合酶beta亚基（DNA-directed RNA polymerase beta subunit）；

  rpoC：RNA聚合酶beta'亚基（DNA-directed RNA polymerase beta'subunit）；

  rps12：核糖体蛋白S12（ribosomal protein S12）；

  rps7：核糖体蛋白S7（ribosomal protein S7）；

  fusA：蛋白延伸因子EF-G（translation elongation factor G）；

  tuf：蛋白延伸因子EF-Tu（translation elongation factor Tu）；

  HP：假拟蛋白（hypothetical protein）。

  根据OY-M株系5'-rplL-rpoB-rpoC-rps12-rps7-fusA-tuf-3'序列（序列结构如图4-7所示）设计引物，采用高保真TaKaRa LA Taq扩增PaWB-sdyz、PaWB-fjfz、LY-fjya1植原体株系 *tuf* 基因及其上游6个基因及间区约12 kb序列，将PCR扩增产物直接测序拼接组装，并对PaWB-sdyz、PaWB-fjfz、LY-fjya1植原体株系持家基因及其编码的氨基酸序列和基因间区序列进行分析。PaWB-sdyz、PaWB-fjfz、LY-fjya1植原体株系及OY-M、AYWB、PAa、SLY、AT共8个植原体株系和1个非固醇甾原体株PG-8A株系的7个基因，以及其对应的氨基酸序列及其6个基因间区的核苷酸序列和氨基酸序列信息如表4-4所示。对 *tuf* 等7个基因及基因间区序列分析发现，PaWB-sdyz、PaWB-fjfz、LY-fjya1、OY-M、AYWB、PAa、SLY植原体株系 *fusA*—*tuf* 基因间区序列长118～130 bp，有明显的启动子结构特征；非固醇甾原体PG-8A株系 *fusA*—*tuf* 基因间区序列长86 bp，有明显的启动子结构特征；AT植原体株系 *fusA*—*tuf* 基因间区序列长65 bp，未检测到完整的启动子结构。PaWB-sdyz、PaWB-fjfz、LY-fjya1、OY-M、PAa、SLY、AT植原体株系 *rps7*—*fusA* 基因间区序列长27～35 bp，基因间区未预测到完整的启动子结构；AYWB株系 *rps7*、*fusA* 为重叠基因；PG-8A株系 *rps7*—*fusA* 基因间区序列长17 bp，亦未检测到完整的启动子结构。PaWB-sdyz、PaWB-fjfz、LY-fjya1、OY-M植原体株系 *rps12*—*rps7* 基因间区序列长为88 bp，有完整的启动子结构；AYWB、PAa、SLY、AT植原体株系 *rps12*—*rps7* 基因间区序列长为51～70 bp，未预测到完整的启动子结构；非固醇甾原体PG-8A株系 *rps12*—*rps7* 基因间区序列长为66 bp，未预测到完整的启动子结构。PaWB-sdyz、PaWB-fjfz、LY-fjya1、OY-M、PAa、SLY、AT植原体株系 *rpoC*—*rps12* 基因间区序列长20～27 bp，基因间区未预测到完整的启动子结构；植原体AT株系 *rpoC*—*rps12* 基因间区序列长193 bp，非固醇甾原体PG-8A株系 *rpoC*—*rps12*

基因间区序列长为260 bp，有完整的启动子保守结构。因非固醇甾原体PG-8A株系*rps12*基因的上游2个基因编码假设蛋白（hypothetical protein），所以PaWB-sdyz、PaWB-fjfz、LY-fjya1、OY-M、PAa、SLY、AT植原体株系的*rplL-rpoB*、*rpoB-rpoC*两个基因间区序列没有再与非固醇甾原体PG-8A株系进行比较分析。PaWB-sdyz、PaWB-fjfz、LY-fjya1、OY-M、PAa、SLY植原体株系*rpoB-rpoC*基因间区序列长2~3 bp，无完整启动子结构；AT株系*rpoB*、*rpoC*基因为重叠基因。PaWB-sdyz、PaWB-fjfz、LY-fjya1、OY-M、PAa、SLY、AT植原体株系*rplL—rpoB*基因间区序列长157~190 bp，预测具有完整的启动子保守区域。植原体和非固醇甾原体*tuf*基因及其上游部分基因结构状况如图4-8所示。关于植原体*fusA—tuf*基因间区，之前我们已对我国不同地区16SrⅠ组植原体株系进行了检测，在泡桐丛枝、苦楝丛枝、莴苣黄化等16SrⅠ组植原体株系*fusA—tuf*基因间区序列发现了4种变异类型，并预测到了完整的启动子结构（于少帅，2016）。4种*fusA—tuf*基因间区序列变异类型的代表株系及其*fusA—tuf*基因间区序列已上传至GenBank数据库（KU563013、KU563014、KU563015、KU563016），其中启动子保守结构已在表4-5列出。根据研究中可能涉及的52个的植原体基因启动子保守区域结构的序列特征，对部分植原体部分基因启动子保守区域不同位置核苷酸种类及出现频率进行了统计分析（表4-6），并在此基础上根据保守区域不同位置核苷酸种类及出现频率推测出可能的植原体基因启动子保守区域序列模式为：

-35区：$T_{90}T_{100}G_{92}T_{75}G_{67}A_{85}$；

-10区：$T_{90}A_{96}T_{92}A_{98}T_{73}T_{90}$。

图4-7　OY-M株系*tuf*基因及其上游基因结构（于少帅，2016）

注：基因间的数字代表基因间区序列长度，单位为bp。

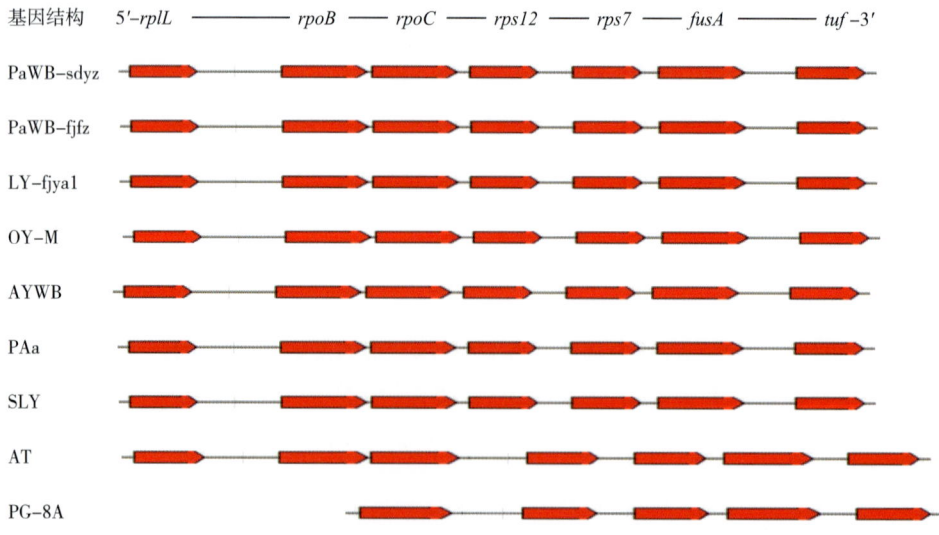

图4-8　不同植原体*tuf*基因及其上游基因结构示意图（于少帅，2016）

表4-4 3株植原体核苷酸序列和编码氨基酸序列及参照株系对应片段长度（于少帅，2016）

株系	扩增序列长度/bp	编码区长度/bp	非编码区长度/bp	氨基酸长度/个
PaWB-sdyz	12 746	12 307	439	4 096
PaWB-fjfz	12 745	12 307	438	4 096
LY-fjya1	12 748	12 307	441	4 096
OY-M	12 745	12 307	438	4 096
AYWB	12 735	12 343	392	4 107
PAa	12 611	12 211	400	4 063
SLY	12 611	12 211	400	4 063
AT	12 835	12 301	534	4 093
PG-8A	4 587	4 158	429	1 382

表4-5 植原体 *rplL—tuf* 基因间区序列相关基因启动子保守区域特征（于少帅，2016）

株系	*rplL-rpoB*		*rpoC-rps12*		*rps12-rps7*		*fusA-tuf*	
	−35	−10	−35	−10	−35	−10	−35	−10
PaWB-sdyz	TTGCAT	TATACC	—	—	ATAAAA	AAAAAT	TTGTGA	TATATT
PaWB-fjfz	TTGCAT	TATACC	—	—	ATAAAA	AAAAAT	TTGTGA	TATATT
LY-fjya1	TTGCAT	TATACC	—	—	ATAAAA	AAAAAT	TTGTAA	TATATT
OY-M	TTGAAT	TATAAC	—	—	ATAAAA	AAAAAT	TTGTGA	TATATT
AYWB	TTGCAT	TATACC	—	—	—	—	TTGTGA	TATTAT
PAa	TTGTAT	TTTAAT	—	—	—	—	TTGATA	TATATT
SLY	TTGTAT	TTTAAT	—	—	—	—	TTGATA	TATATT
AT	ATGATA	AATAAT	TTGACT	TATAAT	—	—	—	—
PG-8A	—	—	TTGACA	TATAAT	—	—	ATGATA	TATTGT

注：研究已发现16SrⅠ组植原体 *fusA—tuf* 基因间区具有完整的启动子结构且存在4种变异类型，31个植原体株系 *fusA—tuf* 基因间区启动子保守区域特征序列与PaWB-sdyz株系一致，分别为泡桐丛枝植原体株系PaWB-jsnj、-jssz1、-jssz2、-bjhr、-hnzz、-hnpy、-hnly、-hncs、-ahhf、-jxnc、-hbbd、-lndl1、-lndl2、-sxty1、-sxty2、-sxxa1、-sxxa2、-sxwn（18个株系）；长春花绿变植原体株系PeV-hnhk（1个株系）；苦楝丛枝植原体株系CWB-hnsy1、-hnsy2、-fjfz1、-fjfz2、-fjfz3、-fjya、-jsnj、-hncs、-jxnc、-gdgz（10个株系）；桑萎缩植原体株系MD-ahhf、-zjca（2个株系）。莴苣黄化植原体株系LY-fjya2株系 *fusA—tuf* 基因间区启动子保守区域特征序列与LY-fjya1株系一致。

表4-6　部分植原体基因启动子保守区域不同位置核苷酸种类及出现频率（于少帅，2016）

保守区域	-35区（5′-3′）						-10区（5′-3′）					
序列特征	A(10)	A(0)	A(8)	A(17)	A(25)	**A(85)**	A(10)	**A(96)**	A(8)	**A(98)**	A(19)	A(0)
	T(90)	**T(100)**	T(0)	**T(75)**	T(6)	T(15)	**T(90)**	T(4)	**T(92)**	T(2)	**T(73)**	**T(90)**
	G(0)	G(0)	**G(92)**	G(0)	**G(67)**	G(0)	G(0)	G(0)	G(0)	G(0)	G(0)	G(0)
	C(0)	C(0)	C(0)	C(8)	C(2)	C(0)	C(0)	C(0)	C(0)	C(0)	C(8)	C(10)

注：-35区、-10区序列不同位置出现的核苷酸频率在碱基后括号中标出（%）；统计的植原体基因启动子个数$n=52$。

启动子的结构和序列变异对启动子的活性及其相关基因的表达影响较大。Du等（2014）发现猪乳糖酶（Pig lactase）基因启动子和增强子中存在较丰富的单核苷酸多态性，这些调控序列的碱基变异导致基因表达水平发生显著差异。Zhang等（2014）研究表明，人类的T-box转录因子编码基因*TBX*1调控序列碱基发生变异，将改变*TBX*1基因的表达水平，而*TBX*1基因表达水平的改变与人类腹股沟疝（inguinal hernia）疾病的产生有一定关系。Ishii等（2009）用启动子软件对植原体质粒序列上的启动子进行了预测和分析，推断ORF3编码蛋白在变异株系OY-NIM中不表达可能是由于启动子的缺失或突变造成的；而且质粒和染色体DNA上不同基因的启动子在-35位和-10位的特征序列核苷酸都有差异。Ishii等（2009）对洋葱黄化植原体野生株系OY-M和其变异非虫传株系OY-NIM质粒ORF1、ORF2基因启动子保守区域的-35区和-10区的特异序列进行了分析，并与OY-M株系16S rRNA、*Amp*基因启动子保守区域的-35区和-10区的特异序列对比分析，各启动子保守区域特征序列如下：

OY-M ORF1：-35区TTCAAT，-10区TATTGA；

OY-M ORF2：-35区TTTATA，-10区TTAATT；

OY-NIM ORF1：-35区CTTAAT，-10区TATTGA；

OY-M 16S rDNA：-35区TTGAAA，-10区TATAAT；

OY-M *Amp*：-35区TTGTTA，-10区TATAAT。

由此发现植原体不同株系、不同功能基因启动子保守区域-35区和-10区的特征序列有所差异。但是，由于统计分析研究中所涉及的启动子数量有限，难以较为系统、全面地揭示植原体启动子保守区域的序列特征。因此，本研究所得到的植原体启动子保守区域可能的序列模式在植原体不同基因启动子中是否具有普遍性和代表性，还有待于后期通过生物信息学方法对大量植原体启动子保守区域进行验证分析。

大肠杆菌基因启动子-35区、-10区序列模式为：

-35区：$T_{85}T_{83}G_{81}A_{61}C_{69}A_{52}$；

-10区：$T_{89}A_{89}T_{50}A_{65}A_{100}$。

由此可见，植原体与大肠杆菌的启动子保守区域结构保守序列碱基存在一定的差异，-35区存在2个碱基差异，碱基T颠换为A，碱基G颠换为C；-10区有1个碱基的差异，碱基T颠换为A。植原体不同株系不同功能基因启动子保守区域序列及其与大肠杆菌启动子保守区域序列模式、非固醇甾原体部分相同基因启动子保守区域序列存在的碱基差异，可能影响不同细菌、不同基因在细菌不同生活阶段、不同环境下的表达及其表达效率，从而达到适应各自不同生态环境的目的（朱玉贤等，2002；Du et al.，2014；Zhang et al.，2014）。

关于植原体启动子分子特征的研究较少。在启动子定性研究的基础上，可以将植原体启动子片段与 *lacZ*（魏云林等，2008；郑庆云等，2013）、*cat*（Palmano et al.，2001）、*GUS*（曹庆银等，2002；蒋国凤等，2014）、*gfp*（周琴等，2004）、碱性蛋白酶（杨春晖，2006；潘皎等，2004）等报告基因连接在受体细胞中表达，通过酶活性检测等方法检测报告基因产物表达量，以及对植原体基因启动子的活性强弱进行精确、定量检测。通过对不同植原体株系、不同植原体基因启动子序列特征的分析，进一步检测、分析植原体基因启动子保守区域的序列特征。在启动子活性精确定量检测的基础上，分析植原体不同启动子类型的结构、功能等分子特征。在此体系的基础上可以通过载体、受体细胞、报告基因等条件的优化探究外界物理条件、化学物质、微生物等对与植原体生长繁殖、致病性等相关的关键基因启动子表达调控的影响，建立植原体与外界环境互作的精确、定量检测体系，克服植原体不能分离培养的缺点，从而筛选出植原体合适的药物靶标序列或其他抗植原体物质，探究植原体关键基因的表达调控特征。

参考文献

曹庆银，陆海，傅学奇，等，2002. 银杏木质部特异定位表达基因启动子在转基因烟草中的功能研究[J]. 北京林业大学学报，24（1）：1-4.

蒋国凤，吴秋菊，梁晓夏，等，2014. 十字花科黑腐病Ⅲ型效应物基因avrACxcc8004推测的启动子区[J]. 微生物学报，54（2）：159-166.

兰婧秋，秦跟基，2021. Class Ⅱ TCP转录因子的主要功能和分子调控机制[J]. 中国科学（生命科学），51（11）：1542-1557.

李继东，陈鹏，倪静，等，2019. 植原体致病分子机理研究进展[J]. 园艺学报，46（9）：1691-1700.

潘皎，张义正，2004. 枯草芽孢杆菌基因启动子的分离与鉴定[J]. 微生物学报，44（4）：

457-460.

魏云林, 季秀玲, 林连兵, 等, 2008. 低温菌启动子分析及异源蛋白高效表达[J]. 生物工程学报, 24 (3): 415-422.

杨春晖, 2006. 短小芽孢杆菌碱性蛋白酶基因启动子的功能研究[D]. 成都: 四川大学.

于少帅, 2016. 植原体 *tuf* 基因启动子分子特征和枣树抗植原体物质研究[D]. 北京: 中国林业科学研究院.

于少帅, 李永, 任争光, 等, 2017. 多位点序列分析揭示我国16SrⅠ组植原体不同株系间遗传变异和系统发育关系[J]. 林业科学, 53 (3): 105-118.

于少帅, 林彩丽, 潘皎, 等, 2016. 泡桐丛枝和枣疯病植原体 *tuf* 基因上游序列结构、功能和遗传变异比较分析[J]. 微生物学通报, 43 (5): 1060-1069.

于少帅, 林彩丽, 王圣洁, 等, 2018. 植原体 *tuf* 基因与其上游部分基因结构和相关基因启动子保守区域特征及活性分析[J]. 生物多样性, 26 (7): 738-748.

于少帅, 徐启聪, 林彩丽, 等, 2016. 植原体遗传多样性研究现状与展望[J]. 生物多样性, 24 (2): 205-215.

赵征慧, 熊鹂, 沈春伟, 等, 2014. 2个适于水稻条斑病菌致病相关基因转录表达分析的启动子探针载体的构建[J]. 植物病理学报, 44 (5): 504-511.

郑庆云, 王冠男, 张喆, 等, 2013. 芽胞外壁基质组成蛋白的编码基因启动子P*exsY*指导的 *cry1Ac* 基因表达[J]. 微生物学报, 54 (10): 1138-1145.

周琴, 孙明, 喻子牛, 2004. 利用绿色荧光蛋白基因 *gfp* 研究芽孢杆菌的启动子活性[J]. 微生物学报, 44 (4): 543-546.

朱玉贤, 李毅, 2002. 现代分子生物学[M]. 2版. 北京: 高等教育出版社.

ANDERSEN M T, LIEFTING L W, HAVUKKALA I, et al., 2013. Comparison of the complete genome sequence of two closely related isolates of 'Candidatus Phytoplasma australiense' reveals genome plasticity[J]. BMC Genomics, 14: 529.

ARSENIJEVIC S, TOPISIROVIC L, 2000. Molecular analysis of mutated *Lactobacillus acidophilus* promoter-like sequence P15[J]. Canadian Journal of Microbiology, 46 (10): 938-945.

BAI X, ZHANG J, EWING A, et al., 2006. Living with genome instability: the adaptation of phytoplasmas to diverse environments of their insect and plant hosts[J]. Journal of Bacteriology, 188 (10): 3682-3696.

CHANG S, TAN C, WU C, et al., 2018. Alterations of plant architecture and phase transition by the phytoplasma virulence factor SAP11[J]. Journal of Experimental Botany, 69 (22): 5389-5401.

CHEN P, CHEN L, YE X, et al., 2022. Phytoplasma effector Zaofeng6 induces shoot proliferation by decreasing the expression of ZjTCP7 in *Ziziphus jujube*[J]. Horticulture Research, 9: uhab032.

CHEN W, LI Y, WANG Q, et al., 2014. Comparative genome analysis of wheat blue dwarf phytoplasma, an obligate pathogen that causes wheat blue dwarf disease in China[J]. PLoS One, 9: e96436.

CHUNG W C, CHEN L L, LO W S, et al., 2013. Comparative analysis of the peanut witches'-broom phytoplasma genome reveals horizontal transfer of potential mobile units and effectors[J]. PLoS ONE, 8（4）: e62770.

DOREE S M, MULKS M H, 2001. Identification of an *Actinobacillus pleuropneumoiae* consensus promoter structure[J]. Journal of Bacteriology, 183（6）: 1983-1989.

DU H T, ZHU H Y, WANG J M, et al., 2014. Single-nucleotide polymorphisms and activity analysis of the promoter and enhancer of the pig lactase gene[J]. Gene, 545（1）: 56-60.

FIRRAO G, MARTINI M, ERMACORA P, et al., 2013. Genome wide sequence analysis grants unbiased definition of species boundaries in "*Candidatus* Phytoplasma"[J]. Systematic and Applied Microbiology, 36（8）: 539-548.

HOSHI A, OSHIMA K, KAKIZAWA S, et al., 2009. A unique virulence factor for proliferation and dwarfism in plants identified from a phytopathognic bacterium[J]. Proceeding of the National Academy of Sciences of the United States of America, 106（15）: 6416-6421.

HUANG W, MACLEAN A M, SUGIO A, et al., 2021. Parasitic modulation of host development by ubiquitin-independent protein degradation[J]. Cell, 184: 5201-5214.

ISHII Y, KAKIZAWA S, HOSHI A, et al., 2009. In the non-insect-transmissible line of onion yellows phytoplasma（OY-NIM）, the plasmid-encoded transmembrane protein ORF3 lacks the major promoter region[J]. Microbiology, 155（6）: 2058-2067.

KAKIZAWA S, OSHIMA K, NISHIGAWA H, et al., 2004. Secretion of immune dominant membrane protein from onion yellows phytoplasma through the Sec protein-translocation system in *Escherichia coli*[J]. Microbiology, 150: 135.

KITAZAWA Y, IWABUCHI N, HIMENO M, et al., 2017. Phytoplasma-conserved phyllogen proteins induce phyllody across the plantae by degrading floral MADS domain proteins[J]. Journal of Experimental Botany, 68（11）: 2799-2811.

KUBE M, SCHNEIDER B, KUHL H, et al., 2008. The linear chromosome of the plant-pathogenic mycoplasma '*Candidatus* Phytoplasma mali'[J]. BMC Genomics, 9: 306.

MACLEAN A, ORLOVSKIS Z, KOWITWANICH K, et al., 2014. Phytoplasma effector SAP54 hijacks plant reproduction by degrading MADS-box proteins and promotes insect colonization in a RAD23-dependent manner[J]. PLoS Biology, 12（4）: e1001835.

MACLEAN A, SUGIO A, MAKAROVA O, et al., 2011. Phytoplasma effector SAP54 induces indeterminate leaf-like flower development in *Arabidopsis* plants[J]. Plant Physiology, 154 (2): 831-841.

MAEJIMA K, IWAI R, HIMENO M, et al., 2014. Recognition of floral homeotic MADS domain transcription factors by a phytoplasmal effector, phyllogen, induces phyllody[J]. Plant Journal, 78 (4): 541-554.

MIYATA S, FURUKI K, OSHIMA K, et al., 2002a. Complete nucleotide sequence of the *S10-spc* operon of phytoplasma: gene organization and genetic code resemble those of *Bacillus subtilis*[J] DNA and Cell Biology, 21 (7): 527-534.

MIYATA S, FURUKI K, SAWAYANAGI T, et al., 2002b. Gene arrangement and sequence of *str* operon of phytoplasma resemble those of *Bacillus* more than those of *Mycoplasma*[J]. Journal of General Plant Pathology, 68 (68): 62-67.

MUSIC M, SAMARZIJA I, HOGENHOUT S, et al., 2019. The genome of 'Candidatus Phytoplasma solani' strain SA-1 is highly dynamic and prone to adopting foreign sequences[J]. Systematic and Applied Microbiology, 42 (2): 117-127.

ORLOVSKIS Z, CANALE M C, HARYONO M, et al., 2017. A few sequence polymorphisms among isolates of Maize bushy stunt phytoplasma associate with organ proliferation symptoms of infected maize plants[J]. Annals of Botany, 119: 869-884.

OSHIMA K, KAKIZAWA S, NISHIGAWA H, et al., 2004. Reductive evolution suggested from the complete genome sequence of a plant-pathogenic phytoplasma[J]. Nature Genetics, 36 (1): 27-29.

PALMANO S, KIRKPATRICK B C, FIRRAO G, 2001. Expression of chloramphenicol acetyltransferase in *Bacillus subtilis* under the control of a phytoplasma promoter[J]. FEMS Microbiology Letters, 199 (2): 177-179.

POST L E, NOMURA M, 1980. DNA sequences from the *str* operon of *Escherichia coli*[J]. The Journal of Biological Chemistry, 255 (10): 4660-4666.

SANANGELANTONI A M, TIBONI O, 1993. The chromosomal location of genes for elongation factor Tu and ribosomal protein S10 in the cyanobacterium *Spirulina platensis* provides clues to the ancestral organization of the *str* and *S10* operons in prokaryotes[J]. Journal of General Microbiology, 139 (11): 2579-2584.

SUGIO A, KINGDOM H N, MACLEAN A M, et al., 2011a. Phytoplasma protein effector SAP11 enhances insect vector reproduction by manipulating plant development and defense hormone biosynthesis[J]. Proceedings of the National Academy of Sciences of the United State of America, 108 (48): 1254-1263.

SUGIO A, MACLEAN A M, KINGDOM H N, et al., 2011b. Diverse targets of phytoplasma effectors: from plant development to defense against insects[J]. Annual Review of Phytopathology, 49(1): 175-195.

TAN C, LI C, TSAO N, et al., 2016. Phytoplasma SAP11 alters 3-isobutyl-2-methoxyprazine biosynthesis in *Nicotaina benthamiana* by suppressing NbOMT1[J]. Journal of Experimental Botany, 67(14): 4415-4425.

TRAN-NGUYEN L T, KUBE M, SCHNEIDER B, et al., 2008. Comparative genome analysis of 'Candidatus Phytoplasma australiense' (subgroup *tuf* Australia I; rp-a) and 'Ca. Phytoplasma asteris' strains OY-M and AY-WB[J]. Journal of Bacteriology, 190(11): 3979-3991.

WANG J, SONG L, JIAO Q, et al., 2018a. Comparative genome analysis of jujube witches'-broom phytoplasma, an obligate pathogen that causes jujube witches'-broom disease[J]. BMC Genomics, 19(1): 689.

WANG N, LI Y, CHEN W, et al., 2018b. Identification of wheat blue dwarf phytoplasma effectors targeting plant proliferation and defense responses[J]. Plant Pathology, 67: 603-609.

WANG N, YANG H, YIN Z, et al., 2018c. Phytoplasma effector SWP1 induces witches'-broom symptom by destabilizing the TCP transcription factor BRANCHED1[J]. Molecular Plant Pathology, 19(12): 2623-2634.

ZHANG Y, HAN Q L, LI C Y, et al., 2014. Genetic analysis of the *TBX1* gene promoter in indirect inguinal hernia[J]. Gene, 535(2): 290-293.

ZURAWSKI G, ZURAWSKI S M, 1985. Structure of the *Escherichia coli S10* ribosomal protein operon[J]. Nucleic Acids Research, 13(12): 4521-4526.

第五章

植原体病害多样性

一、林木植物

桉树

桉树又名尤加利树，是桃金娘科（Myrtaceae）桉属（*Eucalyptus*）树种的统称，包括乔木或灌木类型。已知的600多种桉树中，绝大多数的原产地在澳大利亚，我国引种桉树种类接近80种，主要种植在福建、云南、广东、广西、海南等省（区）。桉树是世界三大速生树种之一，其适应性强，用途广泛，经济价值高，是十分难得的短周期工业用材树种。

发生历史

桉树植原体病害在印度、伊朗、中国、巴西、苏丹、叙利亚、意大利等国均有发生报道。1982年，我国首次在广东桉树上发现桉树植原体病害，感病品种有10余种，其中赤桉（*E. camaldulensis*）、斜纹胶桉（*E. kirtoniana*）和刚果12号桉（*E.* 12ABL）发病较重。

田间症状

桉树感染植原体后，主要表现为节间缩短、侧芽过度生长，进而形成典型的丛枝状。新长出的叶片黄化、卷曲、变小、边缘干枯，病叶易脱落，主脉呈不对称的褐红色。植株生长受阻，严重时病株顶枯直至死亡。

病原及传播方式

16SrⅠ组（Aster yellows group）、16SrⅡ组（Peanut witches'-broom group）和16SrⅫ组（Stolbur group）。可通过嫁接传播。

澳大利亚坚果

澳大利亚坚果是山龙眼科（Proteaceae）澳洲坚果属（*Macadamia*）物种的统称，俗称澳洲坚果，分布于澳大利亚、新喀里多尼亚、苏拉威西岛和马达加斯加的热带雨林中。

在我国，澳大利亚坚果主要栽培于云南、广东、台湾等省，既可作为坚果，也可作为治疗药物，具有很高的经济价值。

发生历史

澳大利亚坚果植原体病害在古巴、美国和我国的云南有发生报道。我国于2021年在云南德宏州的芒市、梁河、盈江、陇川等地发现澳大利亚坚果植原体病害，部分调查地点的发病率超过40%。

田间症状

植株顶芽生长受抑制，导致侧芽丛生呈丛枝状，叶片黄化，叶片变硬，花变叶，植株长势衰退。

病原

16SrⅠ组（Aster yellows group）。

槟榔

槟榔（*Areca catechu*）又名宾门、螺果、仁频、仁榔，是棕榈科（Arecaceae）槟榔属的多年生常绿乔木。槟榔原产于马来西亚，目前主要种植在南亚、东南亚和东亚的部分国家，美洲、东非等部分国家和南太平洋岛国也有少量分布。我国槟榔种植主要集中在海南和台湾两省，广东、广西、云南、福建等省（区）亦有少量种植。槟榔不仅是一种咀嚼嗜好品，还是重要的中药材。

发生历史

槟榔黄化病由植原体引起，是一种毁灭性传染病害。1914年，槟榔黄化病首次发现于印度的喀拉拉邦，目前该病害在印度、中国和斯里兰卡3个国家有发生报道，严重危害当地的槟榔产业。1981年，我国首次在海南省屯昌县发现黄化病。由于海南槟榔种植规模不断扩大，伴随着种苗调运、自然传播等因素，黄化病在海南槟榔种植区逐渐蔓延，导致大量处于旺果期的槟榔树大规模地减产，甚至绝收。

田间症状

槟榔黄化病在田间有2种症状表现，即黄化型和束顶型。在我国，槟榔黄化病绝大多数表现为黄化型，发病初期，槟榔树倒数第2~4张羽状叶片叶缘1/4开始出现黄化，黄色和绿色区域存在明显的分界。随着叶片黄化症状逐年加重，冠幅变小。感病树抽生的花穗短小，果实常提前脱落，少量存留的果实品质差，病叶叶鞘基部的小花苞呈水渍状败坏。植株一般在表现黄化症状后5~7年枯顶死亡；少部分发病槟榔树表现为束顶型，顶部叶片显著缩小，呈束顶状，节间缩短，花穗枯萎无法结果，病叶叶鞘基部的小花苞呈水渍状败坏（图5-1）。

病原及传播方式

16SrⅠ组（Aster yellows group）、16SrⅡ组（Peanut witches'-broom group）、16SrⅪ

组（Rice yellow dwarf group）、16Sr XIV组（Bermudagrass white leaf group）和16Sr XXXII组（Malaysian periwinkle virescence group）。黄化病的远距离传播可能是通过带毒种苗，近距离传播可能通过甘蔗长袖蜡蝉（*Zoraida pterophoroides* Westwood）等刺吸式口器的昆虫传播。

图5-1　槟榔植原体病害

草海桐

草海桐（*Scaevola taccada*）是草海桐科（Goodeniaceae）草海桐属的灌木，我国台湾、福建、广东、广西等地均有栽培。草海桐植原体病害目前仅在阿曼和我国的台湾有发生报道。草海桐感染植原体后，表现出节间缩短、小叶丛生、叶片黄化、丛枝。已发现16Sr II组植原体的侵染。

车桑子

车桑子（*Dodonaea viscosa*）又名明油子、坡柳，是无患子科（Sapindaceae）车桑子属的灌木或小乔木，分布于热带、亚热带和部分温带地区。车桑子常生于干旱山坡、旷地或海边的沙土上，是一种良好的固沙保土树种。

发生历史

植原体侵染引起的车桑子丛枝病在1995年报道于美国夏威夷，随后中国、沙特阿拉伯、伊拉克和埃及等国家也有发生报道。我国仅在四川攀枝花发现车桑子丛枝病。

田间症状

植株生长受到抑制，腋芽和不定芽大量生长，形成丛枝和矮化，叶片褪绿、黄化、变小、皱缩。

病原及传播方式

16SrⅠ组（Aster yellows group）和16SrⅩⅣ组（Bermudagrass white leaf group）。可通过嫁接传播。

滇朴

滇朴（*Celtis tetrandra*）又名四蕊朴、昆明朴、西藏朴、石朴，是大麻科（Cannabaceae）朴属的乔木，分布于不丹、越南、印度、尼泊尔、缅甸和中国等国家。滇朴植原体病害仅在我国云南有发生报道。滇朴幼树易染病，发病严重的植株主干及侧枝上遍布几十个成团簇生的不定芽，这些簇生的不定芽往往不能顺利萌发，易受冻害或日灼影响而枯死。已发现16SrⅠ组植原体的侵染。

海枣

海枣（*Phoenix dactylifera*）又名枣椰、伊拉克枣、椰枣，是棕榈科（Arecaceae）海枣属的常绿大乔木。海枣起源于现在的伊拉克，树龄可达百年，是干热地区重要果树作物之一。海枣的营养价值高，被称为沙漠面包、绿色金子。目前，加那利群岛、北非、中东、巴基斯坦、印度、墨西哥和美国等地均有栽培。我国福建、广东、广西、云南等省（区）有引种栽培，但主要用于园林观赏。

发生历史

海枣感染植原体后，会引起白色顶枯病、致死性黄化病等。目前，苏丹、沙特阿拉伯、埃及、阿曼、伊朗和我国的海南均有海枣植原体病害的发生报道。

田间症状

白色顶枯病最早发现于苏丹，一般出现在5～8年的幼树上，新长出的矛状叶和较老叶的羽状复叶尖端褪绿。叶片上与叶脉平行的方向出现白色区域，并逐渐坏死，呈现枯萎的条纹状。整个树冠会迅速从绿色变为白色。植株在出现症状后6～12个月内就会死亡。

致死性黄化病的症状首先表现为树冠下部1/3的大部分叶片坏死，分枝异常。未坏死叶片上出现黄色条纹，在发病后期，黄化会覆盖树冠上所有的叶片，最终导致植株死亡（图5-2）。

病原

16SrⅠ组（Aster yellows group）、16SrⅡ组（Peanut witches'-broom group）、16SrⅥ组（Clover proliferation group）、16SrⅦ组（Ash yellows group）、16SrⅣ组（Coconut lethal yellows group）、16Sr XIV组（Bermuda white leaf group）。

图5-2 海枣植原体病害

黄槐

黄槐（*Senna surattensis*）又名黄槐决明，是豆科（Fabaceae）决明属的灌木或小乔木，原产南亚、东南亚及大洋洲，主要分布于热带和亚热带地区。我国广西、广东、福建、台湾等省（区）有栽培，常作为绿篱和园林观赏植物。

发生历史

植原体侵染黄槐后，常引起丛枝病和扁枝病。目前，该类病害仅在我国云南有发生报道。

田间症状

丛枝病表现为小叶，在植株的一个生长点处长出许多幼嫩枝条，呈明显的丛枝状。扁枝病表现为植株茎秆变大变平，嫩枝过度增生，受影响的树木或枝条不开花（图5-3）。

病原

16Sr V组（Elm yellows group）和16Sr XII组（Stolbur group）。

图5-3 黄槐植原体病害（引自Wu et al., 2012）

苦楝

苦楝（*Melia azedarach*）又名金铃子、川楝子、楝树，是楝科（Meliaceae）楝属的落叶乔木。苦楝原产于东南亚及澳大利亚北部，广布于亚洲热带和亚热带地区，温带地区也有少量栽培。我国黄河以南各省区均有栽培。苦楝生长迅速、根部扎根广阔且深，可有效防止水土流失，不仅是良好的造林树种，还是优良的用材树种。

发生历史

早在1975年，我国浙江、广东、福建和浙江等省的苦楝树上就有发现丛枝病。目前，该病害在阿根廷、巴西、巴拉圭、叙利亚、印度、越南和韩国等国均有发生报道。

田间症状

苦楝树的顶芽和腋芽大量萌发，叶序紊乱，形成层叠徒长的纤弱枝叶，顶梢病叶簇生成团。枝条节间缩短，叶片变小变细、常伴有黄化症状。丛簇症状往往从个别枝条逐渐蔓延至所有枝条。植株一般先从顶梢开始枯死，小树需经历1～3年，大树则需3～5年，最终全树枯死（图5-4）。

病原和传播方式

16SrⅠ组（Aster yellows group）、16SrⅢ组（X-disease group）、16SrⅥ组（Clover

图5-5　蛇藨筋原体病害

松树

松树是松科（Pinaceae）松属（*Pinus*）植物的统称，有110余种，大多为常绿乔木，广泛分布于北半球。松树用途广泛，不仅是森林更新、造林、绿化的重要树种，还是用材树种。

发生历史

由植原体引起的松树丛枝病在美国、波兰、西班牙、伊朗、立陶宛、中国、法国等国均有报道。2022年在我国四川攀枝花发现该病害，在调查地点，超过9%的松树感染了丛枝病。

田间症状

针叶呈黄色或红色，部分针叶枯萎。一些侧芽成簇生长，少数萌发针叶。

病原

16SrⅠ组（Aster yellows group）、16SrⅥ组（Clover proliferation group）、16SrⅩ组（Apple proliferation group）和16SrⅩⅪ组（Pine shoot proliferation group）。

喜树

喜树（*Camptotheca acuminata*）又名千丈树、旱莲木、薄叶喜树，是蓝果树科（Nyssaceae）喜树属的落叶乔木。喜树植原体病害仅在我国贵州和云南有发生报道。喜树感染植原体后，植株矮化，个别枝条的叶片明显变小，叶片颜色变浅或黄化，腋芽激增，枝芽丛生。发病严重时，枝条枯死。已发现16SrⅩⅩⅫ组（Malaysian periwinkle virescence group）植原体的侵染，可通过小绿叶蝉（*Empoascini* sp.）传播。

细序柳

细序柳（*Salix guebriantiana*）是杨柳科（Salicaceae）柳属的直立灌木，在我国主要分布在云南和四川。细序柳植原体病害仅在我国云南有发生报道。细序柳感染植原体后，枝芽异常激增，小叶，丛枝，花畸形，花绿变。已发现16SrⅠ组（Aster yellows group）植原体的侵染。

相思树

相思树是豆科（Fabaceae）相思树属（*Acacia*）植物的统称，常绿乔木，在我国台湾、海南、福建、广东、广西、云南等地均有种植。相思树生长迅速，耐干旱，为荒山造林、水土保持和沿海防护林的重要树种。

发生历史

相思树植原体病害在印度、美国、澳大利亚、哥伦比亚和我国的海南等地均有发生报道。

田间症状

叶片弯曲，从叶尖开始褪绿，小芽激增成簇生长，呈丛枝状（图5-6）。

病原

16SrⅠ组（Aster yellows group）、16SrⅡ组（Peanut witches'-broom group）和16SrⅦ组（Ash yellows group）

图5-6 相思树植原体病害

橡胶树

橡胶树（*Hevea brasiliensis*）又名三叶橡胶树、巴西橡胶树，是大戟科（Euphorbiaceae）橡胶树属的大乔木，原产于亚马孙森林。橡胶树是最重要的热带经济作物，主要种植在我国海南、云南和广东等省（区），其树汁是天然橡胶的最主要来源。

发生历史

橡胶树丛枝病又称为扁枝病，由植原体引起，1959年首次报道于马来西亚。橡胶树丛枝病在我国植胶垦区可能发生已久，但从1984年起才陆续有文献报道，海南、云南和广东的植胶区均有发生。

田间症状

橡胶树丛枝病在实生苗、芽接苗、幼龄树及已开割的橡胶树上均有发生，感病的橡胶树枝条扁平、畸形，叶片变小或成簇。小苗发生丛枝病后，生长发育不正常。开割树发生丛枝病后，一般会并发橡胶树死皮病（图5-7）。

病原及传播方式

16Sr Ⅰ组（Aster yellows group）和16Sr Ⅱ组（Peanut witches'-broom group）。可通过皮接传播。

图5-7 橡胶树植原体病害

野青树

野青树（*Indigofera suffruticosa*）又名假蓝靛，是豆科（Fabaceae）木蓝属的灌木或亚灌木，原产于热带美洲，我国江苏、浙江、福建、台湾、广东、广西、云南均有栽培。野青树植原体病害仅在我国台湾有发生报道。感染植原体后，植株矮化、小叶、花变叶、花朵增殖。已发现16SrⅡ组（Peanut witches'-broom group）植原体侵染。

樟树

樟树是樟科（Lauraceae）樟属（*Camphora*）植物的统称，常绿乔木，原产于越南，在我国长江以南地区广泛分布。樟树植原体病害仅在我国四川攀枝花有发生报道。感染植原体后，叶片黄化、小叶、丛枝，叶片易脱落。已发现16SrⅠ组（Aster yellows group）植原体侵染。

长梗紫麻

长梗紫麻（*Oreocnide pedunculata*）是荨麻科（Urticaceae）紫麻属的小乔木或灌木，在我国主要分布于台湾。长梗紫麻植原体病害仅在我国的台湾有发生报道。感染植原体后，叶片黄化、丛枝。已发现16SrⅠ组（Aster yellows group）植原体侵染。

竹柏

竹柏（*Nageia nagi*）又名山杉、竹叶柏，是罗汉松科（Podocarpaceae）竹柏属的常绿乔木。叶脉平行似竹叶，材质似杉木。竹柏是有名的景观树与行道树种。

发生历史

由植原体引起的竹柏扁枝病目前仅在海南海口发现，其他地方未见报道。

田间症状

小叶，节间缩短，长出的枝条呈扁平的带状或鸡冠状，染病枝条会逐渐干枯死亡（图5-8）。

病原

16SrⅡ组（Peanut witches'-broom group）和16SrⅫ组（Stolbur group）。

图5-8 竹柏植原体病害

竹

竹是禾本科（Gramineae）竹亚科（subfamily Bambusoideae）植物的统称，多年生高大树状草本植物，一般生长在热带和亚热带地区。在我国，竹的自然分布限于长江流域及其以南各省区，少数种类还可向北延伸至秦岭、汉水及黄河流域各处。

发生历史

由植原体引起的竹丛枝病是竹林中的重要病害之一，目前仅在中国、韩国、菲律宾和印度有发生报道。我国云南、广东、贵州、四川、海南等地均已发现丛枝病，部分竹林的病株率超过50%。

田间症状

最初仅个别枝条发病，小枝丛生，呈典型鸟巢状，丛生的枝条细弱，节间短缩，叶片稍有黄化。病株往往早衰枯死（图5-9）。

病原

16SrⅠ组（Aster yellows group）、16SrⅡ组（Peanut witches'-broom group）和16SrⅧ组（Loofah witches'-broom group）。

图5-9　竹植原体病害

紫檀

紫檀（*Pterocarpus indicus*）又名印度紫檀、花榈木、紫榆，是豆科（Leguminosae）紫檀属的常绿乔木，原产于印度，在我国主要分布于台湾、广东和云南。紫檀树不仅是优良的庭院绿化树种，也是优良的建筑、乐器及家具用材。

发生历史

由植原体引起的紫檀扁枝病目前仅在我国海南有发生报道。

田间症状

植株茎秆变宽、扁平，在扁枝的上部生长点长出许多幼嫩枝条，呈明显的丛枝状（图5-10）。

病原

16SrⅠ组（Aster yellows group）。

图5-10　紫檀植原体病害

二、粮食作物

豆科粮食作物

豆科粮食作物包括大豆、兵豆、赤豆、刀豆、绿豆、芸豆等，其中大豆（*Glycine max*）是最重要的粮食作物之一。大豆又名黄豆，是豆科（Fabaceae）大豆属的一年生草本，约有5 000年的人工栽培史，原产于中国，古称菽，通常被认为是由野生大豆（*Glycine soja* Sieb. et Zucc.）驯化而来，现知约有1 000个栽培品种，广泛栽培于世界各地，我国各地均有栽培，以东北地区最为著名。

发生历史

植原体可危害多种豆科粮食作物，如大豆、兵豆、赤豆、刀豆、绿豆、芸豆等。豆科粮食作物植原体病害在伊朗、印度、泰国、古巴、美国、哥斯达黎加、澳大利亚等国均有发生报道，目前我国台湾、云南、广西等地也有相关报道。

田间症状

植物生长发育不良，枝芽激增呈丛枝状，小叶，黄化，花变叶，花绿变，种子发育不良。

病原

16SrⅠ组（Aster yellows group）、16SrⅡ组（Peanut witches'-broom group）、16SrⅢ组（X-disease group）、16SrⅥ组（Clover proliferation group）、16SrⅨ组（Pigeon pea

全球第六大主食及十亿人口日常生活中的主要热量来源。木薯自1820年引入我国以来，现已在广西、广东、海南、云南等省（区）广泛种植。

发生历史

植原体侵染木薯后可引起丛枝病和蛙皮病。丛枝病在亚洲的越南、柬埔寨、泰国、老挝、菲律宾、印度尼西亚、中国，以及美洲的古巴、巴西、委内瑞拉、墨西哥和秘鲁等国均有发生报道；蛙皮病在美洲的哥伦比亚、巴西、委内瑞拉、秘鲁、哥斯达黎加和巴拿马等国危害严重，我国尚无报道。

田间症状

木薯丛枝病：叶片小而且薄、黄化、扭曲变形，叶序紊乱，茎秆上腋芽大量萌发、节间缩短。茎和根的维管束组织褐变，植株矮化，结薯少或不结薯，严重时死亡。植株矮化，受害严重的植株呈扫帚状。

木薯蛙皮病：发病植株块根数量减少、体积变小、薯块木栓化和表皮腐烂，导致产量和品质降低。受害块根表皮增厚，出现不规则膨胀且容易开裂，常呈现皱褶状，外观类似蟾蜍皮，这是该病害的典型症状。严重时，根系纤维化（或木质化），完全绝收。

病原

16SrⅠ组（Aster yellows group）、16SrⅡ组（Peanut witches'-broom group）、16SrⅢ组（X-disease group）和16SrⅧ组（Loofah witches'-broom group）。

水稻

水稻（*Oryza sativa*）又名稻子、稻谷，是禾本科（Poaceae）稻属的一至二年生草本植物。全世界有半数以上人口以水稻为主食。我国是世界上水稻栽培历史最悠久的国家，南方为主要产稻区，北方部分省份亦有栽培。

发生历史

由植原体侵染引起的水稻橙叶病于1960年首次在泰国发现，随后中国、菲律宾、马来西亚、印度、斯里兰卡、越南、印度尼西亚、柬埔寨等国相继报道了该病害的发生。1978年，我国在云南西双版纳稻区首次发现橙叶病，随后广东、福建、海南、广西、台湾等省相继报道了该病害的发生。20世纪80年代末至90年代初，橙叶病曾在华南局部稻区暴发成灾。

田间症状

橙叶病在水稻的各个生育期都有发生。秧苗期感染，初期症状并不明显，但移栽至大田后，病株相较于健康植株返青延迟，根系发育不良，分蘖盛期开始出现病窝，分蘖减少，叶片短、窄、竖直，严重时全田发黄，在孕穗前枯死或不能抽穗，植株矮小；分蘖期病株基部叶片从叶尖外缘沿叶脉向下呈淡黄到橙黄色条纹，然后向下或从叶缘向中脉扩展，最终导致全叶变橙黄色。随着病势发展，病株中上部叶片逐渐变成橙黄。虽能抽穗，但穗小、粒多不实、米质松脆。重病田呈现大片橙黄色。

病原及传播方式

16Sr XI组（Rice yellow dwarf group）。可通过电光叶蝉（*Inazuma dorsalis*）和黑尾叶蝉（*Nephotettix cinticeps*）传播。

三、油料作物

花生

花生（*Arachis hypogaea*）又名长生果、落花生，是豆科（Fabaceae）落花生属的一年生草本，原产秘鲁和巴西。花生作为重要油料作物，在我国广泛栽培，从北到南均有分布。

发生历史

由植原体引起的花生丛枝病，俗称"花生公"，仅在我国和印度有发生报道。花生丛枝病是我国南方花生产区的主要病害之一，于1952年首次在广东发现。目前，海南、广东、广西、福建、湖南、山东、云南和台湾等地均有发生报道。

田间症状

植株矮化，节间缩短，腋芽大量萌发出丛生枝条，小叶、黄化，花绿变，花变叶，大多数植株不结荚果，严重影响花生的产量（图5-12）。

病原及传播方式

16Sr I组（Aster yellows group）、16Sr II组（Peanut witches'-broom group）和16Sr VI组（Clover proliferation group）。可通过小绿叶蝉（*Empoasca flavescens*）传播。

图5-12　花生植原体病害

芝麻

芝麻（*Sesamum indicum*）又名油麻、脂麻、胡麻，是芝麻科（Pedaliaceae）芝麻属的一年生直立草本，原产于印度。芝麻在汉代时引入我国，栽培极广，既可食用又可作为油料。

发生历史

芝麻植原体病害最早发现于缅甸，在印度、伊朗、伊拉克、以色列、缅甸、苏丹、尼日利亚、坦桑尼亚、巴基斯坦、埃塞俄比亚、泰国、乌干达、墨西哥和我国云南等地均有发生报道，已对许多国家的芝麻生产构成严重威胁。

田间症状

植株矮化，腋芽大量萌发出丛生状枝条，枝条上的叶片细小、皱缩、黄化，花变叶。

病原及传播方式

16Sr Ⅰ 组（Aster yellows group）、16Sr Ⅱ 组（Peanut witches'-broom group）、16Sr Ⅵ 组（Clover proliferation group）和16Sr Ⅸ 组（Pigeon pea witches'-broom group）。可通过菟丝子和网室叶蝉（*Orosius albicinctus*）传播。

四、糖料作物

甘蔗

甘蔗（*Saccharum officinarum*）是禾本科（Poaceae）甘蔗属的多年生高大实心草本植物，主要种植在热带和亚热带地区。我国常年甘蔗种植面积约1 750万亩，主要分布在台湾、福建、广东、海南、广西、四川、云南等地。甘蔗是我国最重要的糖料经济作物，不仅是制造蔗糖的原料，还是制造乙醇的重要原料，也可被用于牲畜饲料。

发生历史

植原体侵染甘蔗可引起白叶病、草芽病和黄叶综合症。甘蔗植原体病害于1947年首次在印度被发现记录，我国于1958年在台湾南部的台南、高雄和屏东发现该类病害，目前广西、广东、云南、福建、海南产区也相继报道。甘蔗植原体病害在世界甘蔗种植国均有发生，是一类严重危害甘蔗产区的病害，如在泰国，甘蔗植原体病害每年造成3 000万～4 000万美元的损失。

田间症状

甘蔗白叶病：叶片上先出现白色或黄色的线条，叶质柔软、白化，分蘖明显增多，株高、茎径、成茎率和单茎重明显降低。病情发展的不同阶段，叶片会呈现不同的白化，有淡绿色、淡黄色、黄加白、淡白加黄、乳白色以及最后的纯白色，植株矮化。

甘蔗草芽病：在上一年的地下残茬基部周围长满绿色草芽，受感染的甘蔗无法成熟。

甘蔗黄叶综合症：一般发生在生长后期，上部叶片中脉首先黄化，并向两侧扩展，中脉下表皮为鲜黄色，上表皮仍是正常的白色或绿白色，部分甘蔗品种的叶片中脉两侧变为红褐色（图5-13）。

病原及传播方式

16SrⅠ组（Aster yellows group）、16SrⅡ组（Peanut witches'-broom group）、16SrⅪ组（Rice yellow dwarf group）和16SrⅫ组（Stolbur group）。可通过种苗、嫁接及细针叶蝉（*Matsumuratettix hiroglyphicus*）和条纹闭颜叶蝉（*Yamatotettix flavovittatus*）传播。

图5-13 甘蔗植原体病害

五、纤维作物

剑麻

剑麻（*Agave sisalana*）又名凤尾兰、菠萝麻，是天门冬科（Asparagaceae）龙舌兰属的多年生植物，原产于墨西哥。我国作为剑麻主产国，产区遍布于广西、广东、海南等省（区）。剑麻是当今世界用量最大，应用范围最广的一种硬质纤维，是国防、捕捞、交通运输、冶金等的重要材料。

发生历史

植原体侵染剑麻引起紫色卷叶病，目前该病害仅在我国有发生报道。2001年11月，海南省昌江青坎农场麻区零星出现了带有紫色卷叶病的剑麻，到2002年4月该病在全场快速蔓延，当年的麻田产量急剧减少。2003年末，紫色卷叶病在广东湛江的剑麻地被发现，部分地块的发病率达到了60%，到2014年时已经蔓延到了广西的剑麻田。目前，紫色卷叶病在我国发生面积约8 400 hm^2，在一些麻田，发病率甚至超过了70%，严重影响剑麻产量。

田间症状

发病初期，剑麻叶片的叶尖边缘处出现紫红色，并且叶尖向内卷曲。随着病情的发展，卷曲现象向下延伸到叶片的中间，且叶片上会出现褪绿黄斑，在卷曲叶片部位的边缘出现紫红色，叶片整体干枯变成黑色并脱落，导致剑麻根系失水死亡，最终使整株剑麻失去其应有的产值。感染紫色卷叶病的剑麻心轴大多会发生腐烂，即感病初期心轴上会有淡黑色的水渍斑点出现，随着叶片的流失，只剩下干枯易断的剑麻纤维，最终会自行断裂脱落（图5-14）。

病原及传播方式

16SrⅠ组（Aster yellows group）。可能通过新菠萝粉蚧（*Dysmicoccus neobreviepes*）传播。

图5-14　剑麻植原体病害（吴伟怀提供）

山黄麻

山黄麻（*Trema tomentosa*）是榆科（Ulmaceae）山黄麻属植物的统称，小乔木或大灌木，我国贵州、西藏、海南、广西、广东、云南、四川、福建等地均有分布。感染植原体后，山黄麻的叶片褪绿、黄化、卷曲、小叶，枝芽异常增殖，远看呈鸟窝状，同时伴有节间缩短，有些丛枝多而密集且整株枯死（图5-15）。已发现16SrⅠ组（Aster yellows group）和16SrXXXⅡ组（Malaysian periwinkle virescence group）植原体的侵染。

图5-15　山黄麻植原体病害

六、果树作物

番木瓜

番木瓜（*Carica papaya*）又名树冬瓜、满万寿果，是番木瓜科（Caricaceae）番木瓜属的绿软木质小乔木，原产热带的美洲，广泛种植于世界热带和较温暖的亚热带地区，在我国海南、台湾、广东、广西、云南、福建等省（区）广泛栽培。

发生历史

番木瓜植原体病害在大洋洲的澳大利亚，美洲的秘鲁、古巴、墨西哥，亚洲的斯里兰卡、巴西、马来西亚、泰国、菲律宾、阿曼、中国、沙特阿拉伯、以色列，非洲的尼日利亚均有发生报道。我国目前仅在台湾和海南有该病害的发生报道。

田间症状

小叶，长势衰退，束顶，冠腐，黄梢（图5-16）。

病原及传播方式

16SrⅠ组（Aster yellows group）、16SrⅡ组（Peanut witches'-broom group）、16SrⅢ组（X-disease group）、16SrⅧ组（Loofah witches'-broom group）、16SrⅫ组（Stolbur group）、16SrⅩⅢ组（Mexican periwinkle virescence group）、16SrⅩⅣ组（Bermudagrass white leaf group）、16SrⅩⅤ组（Hibiscus witches'-broom group）、16SrⅩⅩ组（Buckthorn witches'-broom group）。可通过*Empoasca papayae*和*Colpoptera rotunda*传播。

图5-16　番木瓜植原体病害（3、4由陈旺提供）

番石榴

番石榴（*Psidium guajava*）又名芭乐、菝仔、花稔，是桃金娘科（Myrtaceae）番石榴属的灌木或小乔木，原产于北美洲墨西哥到南美洲北部，广泛种植于热带和亚热带地区，我国台湾、福建、广东、海南、香港、广西、云南南部及四川南部均有栽培。

发生历史

仅在印度和我国海南的番石榴上发现植原体病害。

田间症状

植物长势衰弱，叶片黄化。

病原

16SrⅡ组（Peanut witches'-broom group）。

柑橘

柑橘是野生及栽培环境中芸香科（Rutaceae）柑橘属（*Citrus*）植物的统称，小乔木，包括柚子、酸橙、柑、柠檬等，可能起源于喜马拉雅山的东南山麓。柑橘是世界上产量最大的水果，广布于世界各地。我国柑橘的种质资源丰富，种植面积广泛，生产栽培已覆盖19个省（区、市）。

发生历史

柑橘植原体病害在亚洲的印度、马来西亚、巴基斯坦、阿联酋、沙特阿拉伯、阿曼、伊朗，美洲的巴西、埃塞俄比亚、尼加拉瓜、古巴、墨西哥、智利，北非的埃及，欧洲的法国均有发生报道。柑橘植原体病害在我国的广东、广西、海南、台湾、福建、湖南、云南等柑橘产区均有发生。

田间症状

嫩芽激增呈丛簇状，小叶，叶片黄化（图5-17）。

病原

16SrⅠ组（Aster yellows group）、16SrⅡ组（Peanut witches'-broom group）、16SrⅢ组（X-disease group）、16SrⅣ组（Coconut lethal yellows group）、16SrⅨ组（Pigeon pea witches'-broom group）、16SrⅤ组（Elm yellows group）、16SrⅥ组（Clover proliferation group）、16SrⅦ组（Ash yellows group）、16SrⅫ组（Stolbur group）、16SrⅩⅣ组（Bermudagrass white leaf group）和16SrⅩⅩⅫ组（Malaysian periwinkle virescence group）。

图5-17 柚子植原体病害

梨

梨是蔷薇科（Rosaceae）梨属（*Pyrus*）植物的统称，落叶乔木或灌木，主要分布在亚洲、欧洲至北非。梨作为重要果树及观赏树，在我国各地普遍栽培。

发生历史

梨植原体病害在世界范围内分布广泛，欧洲的俄罗斯、瑞士、德国、西班牙、斯洛文尼亚、捷克、约旦、土耳其、波黑，美洲的美国、阿根廷、意大利、智利、加拿大、乌拉圭，亚洲的中国、印度，大洋洲的澳大利亚等地均有发生报道。目前，我国仅在台湾的梨树上发现该病害。

田间症状

植株长势衰退，叶片黄化、坏死、变形，枝芽增殖，花败育。

病原及传播方式

16SrⅠ组（Aster yellows group）、16SrⅡ组（Peanut witches'-broom group）、16SrⅥ组（Clover proliferation group）、16SrⅩ组（Apple proliferation group）、16SrⅫ组（Stolbur group）、16SrⅩⅢ组（Mexican periwinkle virescence group）和16SrⅩⅩⅨ组（Cassia witches'-broom group）。可通过嫁接、梨木虱（*Psylla pyeicola*）传播。

杧果

杧果（*Mangifera indica*）又名檬果、芒果、蜜望子，是漆树科（Anacardiaceae）杧果属的大乔木，原产自北印度和马来半岛，是世界第五大水果，在我国云南、广西、广东、福建、台湾等地广泛栽培。

发生历史

杧果植原体病害目前仅在巴基斯坦、印度、沙特阿拉伯和我国的云南有发生报道。

田间症状

花变叶，花芽增生，同时伴随花枝缩短，顶梢枯死，新发枝条新叶黄化、畸形、簇生，严重时仅剩叶脉部分，重病植株生长点死亡，植株生长迟缓，最终导致植株死亡，严重影响杧果产量。

病原

16Sr Ⅰ 组（Aster yellows group）、16Sr Ⅱ 组（Peanut witches'-broom group）和16Sr Ⅴ 组（Elm yellows group）。

香蕉

香蕉是芭蕉科（Musaceae）芭蕉属（*Musa*）的多年生常绿草本植物，起源于南亚次大陆到玻利尼西亚等地的东南亚一带，主要生长在热带、亚热带地区。香蕉在我国有2 000多年的栽培历史，我国可分为海南-雷州半岛、粤西-桂南、珠三角-粤东-闽南和桂西南-滇南4个香蕉优势区域。

发生历史

香蕉植原体病害在巴布亚新几内亚、印度、哥伦比亚、印度尼西亚、所罗门群岛和我国广东有发生报道。

田间症状

植原体感染香蕉后，叶片边缘黄化，假茎中出现不连续的棕色或黑色条纹，束顶症状也可能与植原体相关。

病原

16Sr Ⅰ 组（Aster yellows group）和16Sr Ⅱ 组（Peanut witches'-broom group）。

樱桃

樱桃是蔷薇科（Rosaceae）李属樱亚属（*Prunus* subgen. *Cerasus*）所有物种以及李亚属（*Prunus* subgen. *Prunus*）、稠李亚属（*Prunus* subgen. *Padus*）部分物种的统称，既有观赏品种（樱花），又有食用品种。樱桃在中国的栽培范围广泛。

发生历史

樱桃植原体病害在中国、波兰、捷克、印度、法国、伊朗等国有发生报道。我国山东、云南、贵州、四川、台湾等地的樱桃种植区有该类病害的发生报道。

田间症状

叶片黄化、丛枝、花变叶、花绿变、生长衰退，一般在首次出现绿化症状的1个月后，患病的树木枯萎并最终死亡。

病原及传播方式

16SrⅠ组（Aster yellows group）、16SrⅡ组（Peanut witches'-broom group）、16SrⅢ组（X-disease group）、16SrⅤ组（Elm yellows group）、16SrⅥ组（Clover proliferation group）、16SrⅩ组（Apple proliferation group）、16SrⅫ组（Stolbur group）、16SrⅩⅤ组（Hibiscus witches'-broom group）。可通过叶蝉传播。

枣树

枣树是鼠李科（Rhamnaceae）枣属（*Ziziphus*）植物的统称，落叶或常绿乔木，或藤状灌木，原产于中国。枣在我国具有8 000多年的栽培历史，南北各地都有栽培。

发生历史

由植原体引起的枣疯病是生产上的一类具有毁灭性的病害。中国、日本、朝鲜、韩国、印度、伊朗等国均有枣疯病的发生报道。枣疯病于1950年在我国首次被发现，目前在河南、河北、北京、陕西、四川、重庆、新疆、广东等地均有发生报道。

田间症状

枣疯病的典型症状是枝叶丛生和花器返祖。芽萌发形成发育枝，新枝上的芽又多次萌发，形成细弱、节间短、叶片小而黄的丛生状枝条。花器变为营养器官，花柄延长成枝条，花瓣、萼片和雄蕊肥大、变绿、延长成枝叶，雄蕊全部转化成小枝。病株一般不结果，即使结果，果实品质差。枣树根部不定芽大量萌发，长成一丛丛的短疯枝，出土后枝叶细小、黄绿，日晒后全部焦枯呈刷状，根系逐渐腐烂，最后全株死亡。

病原及传播方式

主要为16SrⅤ组（Elm yellows group），也发现16SrⅠ组（Aster yellows group）植原体的侵染。可通过凹缘菱纹叶蝉（*Hishimonus sellatus*）、中华拟菱纹叶蝉（*Hishimonoides chinensis*）和橙带拟菱纹叶蝉（*Hishimonoides aurifascialis*）传播，也可通过嫁接和根蘖苗传播。

七、蔬菜作物

刺芹

刺芹（*Eryngium foetidum*）又名刺芫荽、假香荽，是伞形科（Apiaceae）刺芹属的二年生或多年生草本，主要分布在美洲、亚洲和非洲的热带、亚热带地区，在我国海南、广东、广西等地有栽培。刺芹植原体病害仅在我国海南有发生报道。刺芹感染植原体后，枝芽过度激增呈丛枝状，过度激增的枝芽上长出的花序变小、小花序聚集在顶端呈簇生状，叶片变小、褪绿、扭曲变形（图5-18）。已发现16SrⅠ组（Aster yellows group）的植原体。

图5-18 刺芹植原体病害

番茄

番茄（*Solanum lycopersicum*）又名番柿、西红柿、狼茄，是茄科（Solanaceae）茄属的一年生草本。番茄原产于中美洲和南美洲，现作为食用蔬果已被全球广泛种植。

发生历史

番茄植原体病害最早报道于澳大利亚，随后美国、意大利、伊朗、印度、埃及、约旦、巴西、中国等多个国家相继报道了该类病害的发生。番茄植原体病害于1982年首次在我国海南被发现，随后在新疆、云南、广东也相继发现该病害。

田间症状

番茄感染植原体后，引起巨芽病和丛枝病。巨芽病表现为花芽连同萼片一起膨大，形成囊状物或顶部有锯齿状开口的漏斗形。这样的芽，不能发育成正常的花，而且新梢停止生长，幼叶明显变小，叶缘向上卷曲。丛枝病病株有时表现出几个分枝矮缩，叶片褪绿，长出不正常小叶，茎变厚，且叶子变扭曲变黄。茎上可能出现气生根，由于节间缩短和叶片发育不良，侧芽和植株顶端强烈增殖，植株整体外观呈丛枝状。通常病株不结果或结出少量木质化小果实。在腋芽顶部丛生若干个不定芽（图5-19）。

病原及传播方式

16SrⅠ组（Aster yellows group）、16SrⅡ组（Peanut witches'-broom group）、16SrⅢ组（X-disease group）、16SrⅤ组（Elm yellows group）、16SrⅥ组（Clover proliferation group）、16SrⅩ组（Apple proliferation group）、16SrⅫ组（Stolbur group）、16SrⅩⅢ组（Mexican periwinkle virescence group）。可通过嫁接、菟丝子及网室叶蝉（*Orosius albicinctus*）传播。

图5-19 番茄植原体病害

(图片引自https://www.plantdiseases.org/tomato-big-bud-tomato-0和https://www.gardeningknowhow.com/edible/vegetables/tomato/big-bud-in-tomatoes.htm)

花椰菜

花椰菜(*Brassica oleracea* var. *botrytis*)又称菜花、花菜、开花菜和椰菜花,是十字花科(Brassicaceae)芸薹属的一种常见蔬菜。花椰菜起源于地中海东部,在我国南北方均有栽培。

发生历史

20世纪80年代初,在意大利,科研人员首次通过电子显微镜在表现花变叶、花绿变的花椰菜中发现了植原体。目前,花椰菜植原体病害已经在欧洲、亚洲和美洲等多个国家有发生报道。2016年,我国首次在云南发现花椰菜植原体病害。

田间症状

植株发育不良,叶片轻度变红,芽增殖,花序畸形,花变叶,花绿变。有的花椰菜植株感染植原体后,茎秆扁平,也可能会坏死。

病原及传播方式

16SrⅠ组(Aster yellows group)、16SrⅡ组(Peanut witches'-broom group)、16SrⅢ组(X-disease group)、16SrⅥ组(Clover proliferation group)、16SrⅦ组(Ash yellows group)、16SrⅩⅢ组(Mexican periwinkle virescence group)、16SrⅩⅣ组(Bermudagrass white leaf group)和16SrⅩⅤ组(Hibiscus witches'-broom group)。可通过菟丝子和*Balclutha hebe*传播。

黄瓜

黄瓜（*Cucumis sativus*）又名青瓜、胡瓜、旱瓜，是葫芦科（Cucurbitaceae）黄瓜属的一年生蔓生或攀援草本蔬菜，原产于印度，现广泛种植于温带和热带地区。黄瓜于西汉时期引入我国，现全国各地普遍栽培。

发生历史

黄瓜植原体病害在中国、印度、伊朗、土耳其、立陶宛、马来西亚和埃及等国有发生报道，在我国仅台湾和江苏有发生报道。

田间症状

节间缩短，枝芽异常增殖呈丛枝状，小叶，花绿变，花变叶，扁茎，不育。

病原

16SrⅠ组（Aster yellows group）、16SrⅡ组（Peanut witches'-broom group）、16SrⅥ组（Clover proliferation group）、16SrⅫ组（Stolbur group）。

豇豆

豇豆（*Vigna unguiculata*）又名长豆、赤豆，是豆科（Fabaceae）豇豆属的一年生缠绕、草质藤本或近直立草本，原产于非洲，在我国各地广泛栽培。

发生历史

豇豆植原体病害在中国、澳大利亚、印度、缅甸、伊朗有发生报道。我国的豇豆植原体病害主要发生在海南、台湾、云南、广东和广西等地。

田间症状

叶片黄化，小叶，枝芽异常增殖呈丛枝状，扁茎（图5-20）。

病原

16SrⅠ组（Aster yellows group）、16SrⅡ组（Peanut witches'-broom group）和16SrⅥ组（Clover proliferation group）。

图5-20　豇豆植原体病害

苦瓜

苦瓜（*Momordica charantia*）又名癞葡萄、凉瓜、癞瓜，是葫芦科（Cucurbitaceae）苦瓜属的一年生攀援状柔弱草本，原产于亚洲，现广泛种植于温带和热带地区。

发生历史

苦瓜植原体病害目前在菲律宾、印度、巴西、韩国、马来西亚、澳大利亚及我国的台湾均有发生报道。

田间症状

枝芽激增呈丛枝状，叶片、花序变小，花变叶，节间缩短。

病原

16SrⅠ组（Aster yellows group）、16SrⅡ组（Peanut witches'-broom group）、16SrⅢ组（X-disease group）、16SrⅥ组（Clover proliferation group）和16SrⅧ组（Loofah witches'-broom group）。

辣椒

辣椒又名茝椒仔、番姜、辣子，是茄科（Solanaceae）辣椒属的一年生草本或灌木，起源于美洲。在哥伦布大交换之后，辣椒开始遍布全球。辣椒属有5个种被人工驯化栽培，即一年生辣椒（*Capsicum annuum*）、灌木状辣椒（*Capsicum frutescens*，即小米辣）、中华辣椒（*Capsicum chinense*，即黄灯笼辣椒）、浆果辣椒（*Capsicum baccatum*）、茸毛辣椒（*Capsicum pubescens*，即紫花椒）。

发生历史

1978年，辣椒植原体病害首次在原南斯拉夫被发现，目前在美洲的美国、墨西哥、古巴、哥斯达黎加，欧洲的意大利、塞尔维亚、西班牙、波黑，亚洲的印度尼西亚、伊朗、印度、日本、土耳其、黎巴嫩，大洋洲的澳大利亚均有发生报道。我国的陕西、北京、新疆、湖北、河南、山东、福建、海南等地均有辣椒植原体病害的发生报道。

田间症状

辣椒若在幼苗期感染植原体，植株矮小，叶片黄化，分枝少，易形成"单杆枪"，不结果或偶尔结1~2枚，侧根较少。成株期感病，叶片黄化、皱缩，叶柄变细伸长，枝芽丛生呈扫帚状，严重时茎秆扁化，花序聚生，花器萎落，鳞片发育成小叶（图5-21）。

病原

16SrⅠ组（Aster yellows group）、16SrⅡ组（Peanut witches'-broom group）、16SrⅢ组（X-disease group）、16SrⅥ组（Clover proliferation group）、16SrⅫ组（Stolbur group）、'*Candidatus* Phytoplasma costaricanum'、16SrⅩⅢ组（Mexican periwinkle virescence group）。

图5-21　辣椒植原体病害

萝卜

萝卜（*Raphanus sativus*）又名菜头、莱菔、蓝花子，是十字花科（Brassicaceae）萝卜属的一年生或二年生草本。萝卜可能起源于亚洲或地中海地区，在世界各地均有种植。

发生历史

萝卜植原体病害目前仅在巴基斯坦及我国的台湾和云南均有发生报道。

田间症状

萝卜感染植原体后，叶片变得细小、颜色变深，花瓣变为绿色叶片状，且大量丛生呈簇状，节间短。

病原

16SrⅡ组（Peanut witches'-broom group）。

南瓜

南瓜（*Cucurbita moschata*）又名北瓜、番南瓜、倭瓜，是葫芦科（Cucurbitaceae）南瓜属的一年生蔓生草本，原产于北美洲，世界各地普遍栽培。

发生历史

南瓜植原体病害在马来西亚、澳大利亚及我国台湾、新疆、山东和内蒙古均有发生报道。

田间症状

节间短，每个节生长出多片小叶，果实发育迟缓或没有果实（图5-22）。

病原

16SrⅠ组（Aster yellows group）、16SrⅡ组（Peanut witches'-broom group）和16SrⅫ组（Stolbur group）。

图5-22　南瓜植原体病害

茄子

茄子（*Solanum melongena*）是茄科（Solanaceae）茄属的草本或亚灌木。茄子原产于南亚，在印度和孟加拉国从有刺或苦涩的野生黄水茄经过驯化后传向世界各地。茄子在世界各国都有栽培，但以亚洲产量最多。

发生历史

茄子植原体病害于1939年首次在印度被发现，目前在亚洲的阿曼、伊朗、印度、孟加拉国、日本和中国，欧洲的俄罗斯，大洋洲的澳大利亚，美洲的巴西，非洲的埃及均有发生报道。

田间症状

巨芽，枝芽过度增殖呈丛簇状，小叶，花变绿，花朵畸形且不育，节间缩短，根系分枝过多。

病原

16SrⅠ组（Aster yellows group）、16SrⅡ组（Peanut witches'-broom group）、16SrⅢ组（X-disease group）、16SrⅤ组（Elm yellows group）、16SrⅥ组（Clover proliferation group）、16SrⅨ组（Pigeon pea witches'-broom group）和16SrⅫ组（Stolbur group）。

蛇瓜

蛇瓜（*Trichosanthes anguina*）又名豆角黄瓜、蛇豆，是葫芦科（Cucurbitaceae）栝楼属的一年生攀援藤本，原产于印度和马来西亚，在我国南北各地均有栽培。

发生历史

蛇瓜植原体病害仅在印度和我国台湾有发生报道。

田间症状

枝芽过度增殖呈丛枝状，小叶、黄化，花变叶，花绿变。

病原

16SrⅠ组（Aster yellows group）和16SrⅡ组（Peanut witches'-broom group）。

丝瓜

丝瓜（*Luffa aegyptiaca*）又名吊瓜，是葫芦科（Cucurbitaceae）丝瓜属的一年生攀援藤本，原产于东南亚，广泛栽培于世界温带、热带地区，在我国南北各地普遍栽培。

发生历史

1974年，丝瓜丛枝植原体病害首次在我国台湾发现。该病害在印度、澳大利亚、墨西哥等国也有发生报道。

田间症状

枝芽过度增殖呈簇状，叶子褪绿，开花早，果实较小。

病原及传播方式

16SrⅠ组（Aster yellows group）、16SrⅡ组（Peanut witches'-broom group）、16SrⅧ组（Loofah witches'-broom group）。可通过嫁接和菟丝子传播。

豌豆

豌豆（*Pisum sativum*）又名荷兰豆、雪豆，是豆科（Fabaceae）豌豆属的一年生攀援草本，原产于地中海地区，是最古老的栽培作物之一。豌豆作为重要的粮食和蔬菜作物，广泛栽培于世界各地。

发生历史

豌豆植原体病害目前在阿曼、波兰、伊朗、印度和我国台湾有发生报道。

田间症状

枝芽过度增殖，花变叶，花绿变，无法正常结荚，或者结荚枯萎，或者结荚后的种子较小（图5-23）。

病原

16SrⅠ组（Aster yellows group）、16SrⅡ组（Peanut witches'-broom group）、16SrⅨ组（Pigeon pea witches'-broom group）和16SrⅫ组（Stolbur group）。

图5-23 豌豆植原体病害（Chiu et al., 2023）

蕹菜

蕹菜（*Ipomoea aquatica*）又名空心菜、藤菜、通菜，是旋花科（Convolvulaceae）番薯属的一年生蔓生草本，原产于东亚，现主要分布于亚洲，在我国主要分布于长江以南地区。

发生历史

蕹菜植原体病害目前仅在我国云南有发生报道。

田间症状

叶片变小，枝芽增殖呈丛枝状。

病原

16SrⅡ组（Peanut witches'-broom group）。

莴苣

莴苣（*Lactuca sativa*）又名生菜、莴菜、春菜，是菊科（Asteraceae）莴苣属的一年生或二年生草本。莴苣原产于地中海地区，因适应性佳而遍及温带、亚热带地区，长久以来被人类培育为世界重要的蔬菜。莴苣在我国各地均有栽培，同时也有野生分布。

发生历史

植原体侵染莴苣后可引起变叶病、丛枝病、黄化病和褪绿心腐病。莴苣植原体病害最早于1989年报道于美国，随后在加拿大、意大利、澳大利亚、中国、伊朗、阿根廷、智利、黎巴嫩等国相继报道，目前我国云南、新疆和福建也有发生报道。

田间症状

莴苣变叶病、丛枝病和黄化病的主要症状为：叶片小、褪绿和变形，枝芽过度激增呈丛枝状，老叶变红，叶柄伸长，花绿变，花变叶，不育，发育迟缓。

莴苣褪绿心腐病主要症状为：心叶首先出现褪绿或褪红，叶色褪绿变白，之后顶部叶片挺直变细拔高，逐渐停止生长，植株矮小，嫩叶基部流胶，最后心腐死株。

病原及传播方式

16SrⅠ组（Aster yellows group）、16SrⅡ组（Peanut witches'-broom group）、16SrⅢ组（X-disease group）和16SrⅨ组（Pigeon pea witches'-broom group）。可通过黑褐环茎叶蝉（*Neoaliturus fenestratus*）传播。

八、香辛料作物

胡椒

胡椒（*Piper nigrum*）是胡椒科（Piperaceae）胡椒属的木质攀援藤本，原产于东南亚，现广植于热带地区，在我国台湾、福建、广东、广西及云南等省（区）均有栽培。胡椒植原体病害仅在印度和我国海南有发生报道。胡椒感染植原体后，叶片黄化和卷曲、花变叶、丛枝、长势缓慢，在症状晚期，胡椒藤变黄变细，严重影响产量（图5-24）。已发现16SrⅠ组（Aster yellows group）植原体的侵染。

图5-24　胡椒植原体病害

罗勒

罗勒（*Ocimum basilicum*）又名兰香、香草、九层塔、光明子，是唇形科（Labiatae）罗勒属的一年生或多年生草本，原产于亚洲和非洲，在我国的中部、南部和东南部均有栽培。罗勒植原体病害在古巴、沙特阿拉伯、印度和中国云南均有发生报道。罗勒感染植原体后，枝芽过度增殖呈丛枝状、小叶、花变叶、茎秆上出现黑色条纹，已发现16SrⅠ组（Aster yellows group）和16SrⅡ组（Peanut witches'-broom group）植原体的侵染。

桂树

桂树是樟科（Lauraceae）樟属（*Cinnamomum*）植物的统称，乔木，在我国热带及亚热带地区广为栽培，其中尤以广西栽培最多。桂树植原体病害仅在越南及我国的海南、四川和云南有发生报道。桂树感染植原体后，叶片黄化，枝芽激增呈丛枝状，小叶，植株的茎、枝、叶柄和静脉上形成小的瘤状突起，瘤状物变长如鱿鱼须（图5-25）。发病植株发育迟缓，导致肉桂产量和质量大幅下降。已发现16SrⅠ组（Aster yellows group）、16SrⅡ组（Peanut witches'-broom group）、16SrⅩⅣ组（Bermudagrass white leaf group）植原体的侵染。

图5-25　桂树植原体病害（Chen et al.，2023）

紫苏

紫苏（*Perilla frutescens*）又名苏子、香荽，是唇形科（Labiatae）紫苏属的一年生直立草本，在我国各地均有栽培，供药用及香料用。紫苏植原体病害仅在我国云南有发生报道。紫苏感染植原体后，枝芽激增呈丛枝状，小叶。已发现16SrⅠ组（Aster yellows group）植原体的侵染。

九、饲料及绿肥作物

短绒野大豆

短绒野大豆（*Glycine tomentella*）又名阔叶大豆、绒毛大豆，是豆科（Fabaceae）大豆属的多年生缠绕或匍匐草本，在我国的台湾、福建、广东等地有种植分布。短绒野大豆植原体病害仅在我国台湾有发生报道，感病植株表现为小叶、丛枝、节间缩短、发育迟缓等症状。已发现16SrⅡ组（Peanut witches'-broom group）植原体的侵染。

柱花草

柱花草（*Stylosanthes guianensis*）是豆科（Fabaceae）笔花豆属的多年生草本或亚灌木，是热带和亚热带地区优良的豆科牧草，在我国的广东、广西、海南、云南、福建和台湾等省（区）广泛栽培。柱花草植原体病害在澳大利亚及我国的云南、海南有发生报道。柱花草感染植原体后，叶片黄化、小叶、枝芽激增呈丛枝状（图5-26）。已发现16SrⅡ组（Peanut witches'-broom group）和16SrⅧ组（Loofah witches'-broom group）植原体的侵染。

图5-26　柱花草植原体病害

灰毛豆

灰毛豆（*Tephrosia purpurea*）又名红花灰叶、假蓝靛、灰叶，是豆科（Fabaceae）灰毛豆属的亚灌木状草本，广布于全世界热带地区，枝叶可做绿肥。灰毛豆植原体病害在伊朗、印度和我国的海南有发生报道。灰毛豆感染植原体后，叶片褪绿、小叶、枝芽激增呈丛枝状、节间缩短（图5-27）。已发现16SrⅡ组（Peanut witches'-broom group）和16SrⅫ组（Stolbur group）植原体的侵染。

图5-27　灰毛豆植原体病害

木豆

木豆（*Cajanus cajan*）又名三叶豆，是豆科（Fabaceae）木豆属的直立灌木，在热带和亚热带地区普遍有栽培，叶可作家畜饲料和绿肥。木豆植原体病害在印度、墨西哥、缅甸、澳大利亚和我国的海南有发生报道。木豆感染植原体后，节间缩短，植株明显矮化，叶片黄化，小叶，枝芽激增呈丛枝状，发育迟缓（图5-28）。已发现16Sr I 组（Aster yellows group）、16Sr II 组（Peanut witches'-broom group）、16Sr VI 组（Clover proliferation group）、16Sr IX 组（Pigeon pea witches'-broom group）和16Sr XIV 组（Bermudagrass white leaf group）植原体的侵染。

图5-28　木豆植原体病害

苜蓿

苜蓿（*Medicago sativa*）是豆科（Fabaceae）苜蓿属的多年生草本，在世界各国有广泛种植，是优良的牧草。苜蓿植原体病害发生广泛，亚洲的中国、印度、阿曼、土耳其、巴基斯坦、伊朗、沙特阿拉伯，欧洲的波兰、意大利、俄罗斯、立陶宛、塞尔维亚，大洋洲的澳大利亚均有发生报道。我国的云南、陕西和新疆有该类病害的发生报道。苜蓿感染植原体后，枝芽激增呈丛枝状，小叶，花变叶（图5-29）。已发现16SrⅠ组（Aster yellows group）、16SrⅡ组（Peanut witches'-broom group）、16SrⅤ组（Elm yellows group）、16SrⅥ组（Clover proliferation group）、16SrⅨ组（Pigeon pea witches'-broom group）和16SrⅫ组（Stolbur group）植原体的侵染。

图5-29 苜蓿植原体病害

山蚂蝗

山蚂蝗是豆科（Fabaceae）山蚂蝗属（*Desmodium*）植物的统称，草本、亚灌木或灌木，是热带和亚热带地区优良的豆科牧草。山蚂蝗植原体病害仅在我国海南和台湾有发生报道。山蚂蝗感染植原体后，枝芽增殖呈丛枝状，小叶，发育迟缓，随着病情的发展，感病枝叶枯死（图5-30）。已发现16SrⅡ组（Peanut witches'-broom group）植原体的侵染。

猪屎豆

猪屎豆（*Crotalaria pallida*）又名黄野百合，是豆科（Fabaceae）猪屎豆属的多年生草本或呈灌木状，主要分布在热带和亚热带地区。猪屎豆植原体病害仅在越

图5-30 山蚂蝗植原体病害

南和我国的海南、云南有发生报道。猪屎豆感染植原体后，枝芽异常激增，小叶，丛枝（图5-31）。已发现16SrⅠ组（Aster yellows group）和16SrⅡ组（Peanut witches'-broom group）植原体的侵染。

图5-31 猪屎豆植原体病害

十、观赏植物

矮牵牛

矮牵牛是茄科（Solanaceae）矮牵牛属（*Petunia*）的一年生草本植物的统称，在世界各国花园中普遍栽培。矮牵牛植原体病害在印度、伊朗、沙特阿拉伯、韩国和我国的云南均有发生报道。矮牵牛感染植原体后，顶芽和腋芽异常增多，发育后的叶片小呈丛枝状，花变叶，花绿变，开花少或不开花，花茎弯曲坏死，茎发育为扁平状，且在片状茎长出很多小叶片，植株发育迟缓、矮化。已发现16SrⅠ组（Aster yellows group）和16SrⅡ组（Peanut witches'-broom group）植原体的侵染。

百合

百合是百合科（Liliaceae）百合属（*Lilium*）一年生草本植物的统称，在我国分布范围广。百合植原体病害在墨西哥、捷克、韩国和我国的云南有发生报道。百合感染植原体后，节间缩短、矮化，主茎变宽扁平，枝芽激增，发育后的叶片小呈丛枝状，花小，花瓣变绿，花柱畸形，败育，幼根死亡。已发现16SrⅠ组（Aster yellows group）和16SrⅫ组（Stolbur group）植原体的侵染。

长春花

长春花（*Catharanthus roseus*）又名日日春、长春海棠，是夹竹桃科（Apocynaceae）

长春花属的多年生草本，原产于非洲东部，现广泛栽培于世界热带和亚热带地区。长春花植原体病害在世界各地长春花栽培区多有发生。长春花感染植原体后，植株矮化，枝芽异常激增呈丛簇状，小叶，叶片黄化，花变叶，花绿变（图5-32）。长春花是植原体研究中常用的模式植物，可以被已知的大多数植原体侵染。在自然条件下，在长春花中已发现16SrⅠ组（Aster yellows group）、16SrⅡ组（Peanut witches'-broom group）、16SrⅢ组（X-disease group）、16SrⅤ组（Elm yellows group）、16SrⅦ组（Ash yellows group）、16SrXXXII组（Malaysian periwinkle virescence group）植原体的侵染。

图5-32 长春花植原体病害

二列黑面神

二列黑面神（*Breynia disticha*）又名雪花木、山漆茎、白雪树，是叶下珠科（Phyllanthaceae）黑面神属的常绿灌木，在我国岭南地区分布较广。二列黑面神植原体病害仅在我国广东有发生报道。小叶是该病害的典型症状。已发现16SrⅥ组（Clover proliferation group）植原体的侵染。

海芋

海芋（*Alocasia odora*）是天南星科（Araceae）海芋属的大型常绿草本植物，主要分布在热带和亚热带地区。海芋植原体病害仅在我国的海南有发生报道。海芋感染植原体后，叶片黄化（图5-33）。已发现16SrⅠ组（Aster yellows group）植原体的侵染。

金合欢

金合欢（*Vachellia farnesiana*）又

图5-33 海芋植原体病害

名牛角花、刺球花、鸭皂树，是豆科（Fabaceae）金合欢属的灌木或小乔木，在我国的台湾、福建、广东、广西、浙江等地有分布。金合欢植原体病害仅在我国云南有发生报道，花变叶是该病害的典型症状。已发现16SrⅡ组（Peanut witches'-broom group）植原体的侵染。

九里香

九里香（*Murraya exotica*）又名月橘、四季青、黄金桂，是芸香科（Rutaceae）九里香属的小乔木，在我国的台湾、福建、广东、海南、广西等地均有分布。九里香植原体病害仅在我国台湾有发生报道。九里香感染植原体后，叶片黄化，小叶和丛枝。已发现16SrⅠ组（Aster yellows group）植原体的侵染。

菊花

菊花（*Chrysanthemum × morifolium*）又名小白菊、滁菊、绿牡丹，是菊科（Asteraceae）菊属的多年生草本，原产于我国和日本。菊花植原体病害仅在印度、埃及、巴西及我国的台湾和内蒙古均有发生报道。菊花感染植原体后，腋芽激增呈丛枝状，小叶，花变叶，开花延迟，花萼和花瓣均变成绿色，绿变后的花朵明显小于正常花朵，植株矮化（图5-34）。已发现16SrⅠ组（Aster yellows group）和16SrⅡ组（Peanut witches'-broom group）植原体的侵染。

图5-34　菊花植原体病害（Tseng et al., 2024）

蓝猪耳

蓝猪耳（*Torenia fournieri*）又名夏堇、兰猪耳，是母草科（Linderniaceae）蝴蝶草属的一年生草本，原产越南，在我国南方广泛栽培。蓝猪耳植原体病害仅在我国台湾有发生报道。蓝猪耳感染植原体后，枝芽过度增殖呈丛枝状，叶片黄化，植株矮化。已发现16SrⅠ组（Aster yellows group）植原体的侵染。

蔓花生

蔓花生（*Arachis duranensis*）又名假花生，是豆科（Fabaceae）落花生属的多年生宿根草本，原产于亚洲热带及南美洲，在我国南方广泛栽培。蔓花生植原体病害仅在我国海南和台湾有发生报道。蔓花生感染植原体后，叶片黄化，叶面积较健康植株大，叶肉变薄，长势衰弱（图5-35）。已发现16SrⅡ组（Peanut witches'-broom group）植原体的侵染。

图5-35　蔓花生植原体病害

千穗谷

千穗谷（*Amaranthus hypochondriacus*）是苋科（Amaranthaceae）苋属的一年生草本植物，原产北美，在我国的云南、内蒙古、河北、四川等地有栽培。千穗谷植原体病害在墨西哥、印度和我国广东有发生报道。千穗谷感染植原体后，花朵畸形，顶生花穗的花序轴变扁平，侧生穗激增呈丛簇状。已发现16SrⅠ组（Aster yellows group）和16SrⅡ组（Peanut witches'-broom group）植原体的侵染。

秋英

秋英（*Cosmos bipinnatus*）又名格桑花、扫地梅、大波斯菊，是菊科（Asteraceae）秋英属的一年生或多年生草本，原产于美洲墨西哥，在我国各地均有分布。秋英植原体病害仅在伊朗和我国台湾有发生报道。秋英感染植原体后，小叶，花变叶，花绿变。已发现16SrⅠ组（Aster yellows group）和16SrⅡ组（Peanut witches'-broom group）植原体的侵染。

洒金榕

洒金榕（*Codiaeum variegatum*）又名变叶木，是大戟科（Euphorbiaceae）变叶木属的灌木或小乔木，原产于亚洲马来半岛至大洋洲，在热带地区广泛栽培。洒金榕植原体病害在乌干达、印度及我国的海南和台湾有发生报道。洒金榕感染植原体后，枝芽异常增殖，发育后的叶片小呈丛簇状。已发现16SrⅠ组（Aster yellows group）和16SrⅡ组（Peanut witches'-broom group）植原体的侵染。

石竹

石竹是石竹科（Caryophyllaceae）石竹属（*Dianthus*）植物的统称，多年生草本，在我国各地广泛栽培。石竹植原体病害在印度、塞尔维亚及我国的云南、台湾和陕西有发生报道。石竹感染植原体后，从花、花状结构及植物的基部产生大量的新芽，花变叶，花绿变，心叶变黄，植株矮化。已发现16SrⅠ组（Aster yellows group）、16SrⅡ组（Peanut witches'-broom group）、16SrⅤ组（Elm yellows group）和16SrⅫ组（Stolbur group）植原体的侵染。

松果菊

松果菊（*Echinacea purpurea*）又名紫锥菊、紫锥花，是菊科（Asteraceae）松果菊属的一年生草本植物，原产于北美洲东部，在世界各地广泛栽培。松果菊植原体病害在澳大利亚、伊朗、捷克、匈牙利、斯洛文尼亚和我国的台湾有发生报道。松果菊感染植原体后，叶片黄化或者变红，丛枝，花变叶，花绿变。已发现16SrⅠ组（Aster yellows group）和16SrⅡ组（Peanut witches'-broom group）植原体的侵染。

仙人掌

仙人掌（*Opuntia dillenii*）是仙人掌科（Cactaceae）仙人掌属的丛生肉质灌木，在我国南方沿海地区常见栽培，亦有野生。仙人掌植原体病害在墨西哥、黎巴嫩、意大利、埃及、土耳其和我国的云南有发生报道。仙人掌感染植原体后，叶状茎变形，茎的侧面和顶端长出许多肉质的圆柱形枝条，花败育，植株矮化（图5-36）。已发现16SrⅠ组（Aster yellows group）、16SrⅡ组（Peanut witches'-broom group）和16SrⅥ组（Clover proliferation group）植原体的侵染。

图5-36 仙人掌植原体病害（Cai et al., 2008）

蟹爪兰

蟹爪兰（*Schlumbergera truncata*）又名螃蟹兰、圣诞仙人掌，是仙人掌科（Cactaceae）仙人指属的肉质灌木，原产于南美的巴西，在我国各地均有栽培。蟹爪兰植原体病害仅在我国云南有发生报道。蟹爪兰感染植原体后，叶状茎两侧丛生很多小的叶状茎，不能开花，植株矮化。已发现16SrⅡ组（Peanut witches'-broom group）植原体的侵染。

一点红

一点红（*Emilia sonchifolia*）又名紫背叶、花古帽、野木耳菜，是菊科（Asteraceae）一点红属的一年生草本，广泛分布于热带和亚热带地区。一点红植原体病害在我国的海南和台湾有发生报道。一点红感染植原体后，节间缩短，叶片褪绿，小叶（图5-37）。已发现16SrⅡ组（Peanut witches'-broom group）植原体的侵染。

图5-37　一点红植原体病害

一品红

一品红（*Euphorbia pulcherrima*）又名圣诞花、老来娇、猩猩木，是大戟科（Euphorbiaceae）大戟属的灌木，原产于美洲，广泛栽培于热带和亚热带地区。一品红植原体病害在芬兰、意大利、日本、韩国、土耳其、加拿大和我国的台湾有发生报道。一品红感染植原体后，叶片变窄，顶芽和腋芽异常增多呈丛生状，顶端优势减弱，花蕾增殖，苞片卷曲，有时也会引起扁茎。已发现16SrⅠ组（Aster yellows group）、16SrⅢ组（X-disease group）和16SrⅫ组（Stolbur group）植原体的侵染。可通过嫁接或菟丝子传播。

紫芳草

紫芳草（*Exacum affine*）是龙胆科（Gentianaceae）藻百年属的一年生草本植物，原产于非洲索科得拉岛，适于盆栽观赏，我国各地均有栽培。紫芳草植原体病害目前仅在我国的台湾有发生报道。紫芳草感染植原体后，叶片黄化，植株矮化。已发现16Sr I 组（Aster yellows group）植原体的侵染。

十一、其他植物

百慕达草

百慕达草（*Cynodon dactylon*）又名狗牙根，是禾本科（Poaceae）狗牙根属的一种多年生低矮草本，原产非洲，广泛分布于热带、亚热带和温带地区。百慕达草植原体病害在印度、缅甸、意大利、阿尔巴尼亚、塞尔维亚、古巴、埃塞俄比亚、沙特阿拉伯、越南、斯里兰卡、老挝、泰国、坦桑尼亚、苏丹、肯尼亚及我国的陕西和台湾均有发生报道。百慕达草感染植原体后，叶片上出现浅绿色至黄色条纹，随着病情的发展叶片完全变黄或变白。已发现16Sr XI组（Rice yellow dwarf group）、16Sr XII组（Stolbur group）、16Sr XIV组（Bermudagrass white leaf group）植原体的侵染。

慈姑

慈姑是泽泻科（Alismataceae）慈姑属（*Sagittaria*）的多年生沼生草本植物的统称。我国除西藏外，各地均有分布。慈姑植原体病害仅在我国广西有发生报道。慈姑感染植原体后，叶缘和叶脉首先发黄，然后叶脉间的叶肉逐渐黄化，严重时整株黄化枯死。已发现16Sr I 组（Aster yellows group）植原体的侵染。

二萼丰花草

二萼丰花草（*Spermacoce exilis*）是茜草科（Rubiaceae）钮扣草属的矮小草本，在我国香港和海南等地广泛分布。二萼丰花草植原体病害仅在我国的海南有发生报道。花变叶是该病害的典型症状（图5-38）。已发现16Sr II组（Peanut witches'-broom group）植原体的侵染。

图5-38　二萼丰花草植原体病害（陈旺提供）

广寄生

广寄生（*Taxillus chinensis*）是桑寄生科（Loranthaceae）钝果寄生属的灌木，在我国主要分布于广西、广东、福建等省（区）。广寄生植原体病害仅在我国的广东有发生报道。广寄生感染植原体后，叶片褪绿，小叶，丛枝，发育迟缓（图5-39）。已发现16Sr V组（Elm yellows group）植原体的侵染。

图5-39　广寄生植原体病害（Xi et al.，2022）

鬼针草

鬼针草（*Bidens pilosa*）是菊科（Asteraceae）鬼针草属的一年生草本，在我国华东、华中、华南、西南各地广泛分布。鬼针草植原体病害在巴西和我国的海南有发生报道。鬼针草感染植原体后，叶片褪绿、卷曲。已发现16Sr I组（Aster yellows group）、16Sr III组（X-disease group）和16Sr IX组（Pigeon pea witches'-broom group）植原体的侵染。

黄花草

黄花草（*Arivela viscosa*）又名臭矢菜、野油菜、黄花菜，是白花菜科（Cleomaceae）黄花草属的一年生直立草本，主要分布在热带和亚热带地区。黄花草植原体病害仅在印度和我国的海南有发生报道。黄花草感染植原体后，枝芽异常激增，叶片细小，枝条分枝角度小，呈扫帚状直立（图5-40）。已发现16Sr II组（Peanut witches'-broom group）植原体的侵染。

图5-40　黄花草植原体病害

黄花稔

黄花稔是锦葵科（Malvaceae）黄花稔属（*Sida*）的直立亚灌木植物的统称，主要分布于热带和亚热带地区。黄花稔植原体病害在澳大利亚、巴西和我国的海南有发生报道。黄花稔感染植原体后，枝芽激增呈丛枝状，小叶，花绿变，长势衰弱（图5-41）。已发现16SrⅡ组（Peanut witches'-broom group）和16SrⅩⅤ组（Hibiscus witches'-broom group）植原体的侵染。

图5-41　黄花稔植原体病害（陈旺提供）

假臭草

假臭草（*Praxelis clematidea*）是菊科（Asteraceae）假臭草属的一年生草本，在我国海南、广东等地广泛分布。假臭草植原体病害在日本、澳大利亚、泰国和我国的海南有发生报道。假臭草感染植原体后，大量茎（芽）生长异常呈丛枝状，叶片细小，花变叶，花器萎落（图5-42）。已发现16SrⅠ组（Aster yellows group）和16SrⅡ组（Peanut witches'-broom group）植原体的侵染。

图5-42　假臭草植原体病害

假马鞭

假马鞭（*Stachytarpheta jamaicensis*）是马鞭草科（Verbenaceae）假马鞭属多年生粗壮草本或亚灌木状，主要分布在热带和亚热带地区。假马鞭植原体病害在牙买加、加纳、印度、科特迪瓦和我国的海南有发生报道。假马鞭感染植原体后，叶片褪绿、上卷，小叶，枝芽异常增殖呈丛簇状，长势衰退（图5-43）。已发现16SrⅡ组（Peanut witches'-broom group）、16SrⅣ组（Coconut lethal yellows group）和16SrⅩⅫ组（Nigerian coconut lethal decline group）植原体的侵染。

图5-43　假马鞭植原体病害（陈旺提供）

绞股蓝

绞股蓝（*Gynostemma pentaphyllum*）又名毛绞股蓝，是葫芦科（Cucurbitaceae）绞股蓝属的草质攀援藤本，在我国陕西南部和长江以南各省区均有分布。绞股蓝植原体病害仅在我国云南有发生报道。绞股蓝感染植原体后，枝芽异常增殖呈丛枝状，小叶。已发现16SrⅠ组（Aster yellows group）植原体的侵染。

菊苣

菊苣（*Cichorium intybus*）是菊科（Asteraceae）菊苣属的多年生草本，在我国广泛分布。菊苣植原体病害在意大利、沙特阿拉伯和我国的云南有发生报道。菊苣感染植原体后，枝芽异常增殖呈丛簇状，小叶，花变叶，花绿变，矮化。已发现16SrⅡ组（Peanut witches'-broom group）、16SrⅨ组（Pigeon pea witches'-broom group）和16SrⅫ组（Stolbur group）植原体的侵染。

鳢肠

鳢肠（*Eclipta prostrata*）又名凉粉草、旱莲草，是葫芦科（Cucurbitaceae）鳢肠属的一年生草本，在热带及亚热带地区广泛分布。鳢肠植原体病害仅在我国的海南和台湾有发生报道。鳢肠感染植原体后，花器官绿化、大量枝芽激增呈丛簇状（图5-44）。已发现16SrⅡ组（Peanut witches'-broom group）植原体的侵染。

图5-44 鳢肠植原体病害（陈旺提供）

马松子

马松子（*Melochia corchorifolia*）又名野路葵，是锦葵科（Malvaceae）马松子属的亚灌木状草本，在我国长江以南各地广泛分布。马松子植原体病害仅在我国海南有发生报道。马松子感染植原体后，叶片褪绿，小叶，丛枝，节间短，花变叶，花绿变（图5-45）。已发现16SrⅠ组（Aster yellows group）和16SrⅡ组（Peanut witches'-broom group）植原体的侵染。

图5-45 马松子植原体病害（陈旺提供）

蔓草虫豆

蔓草虫豆（*Cajanus scarabaeoides*）又名虫豆、白蔓草虫豆，是豆科（Fabaceae）木豆属的蔓生或缠绕状草质藤本，主要分布在热带和亚热带地区。蔓草虫豆植原体病害仅在我国云南有发生报道。蔓草虫豆感染植原体后，植株明显矮化，小叶、黄化，大量细弱的枝条丛生成簇。已发现16SrⅡ组（Peanut witches'-broom group）植原体的侵染。

玫瑰茄

玫瑰茄（*Hibiscus sabdariffa*）又名山茄子、洛神花，是锦葵科（Malvaceae）木槿属一年生直立草本，广泛分布在热带和亚热带地区。玫瑰茄植原体病害仅在印度和我国的台湾有发生报道。玫瑰茄感染植原体后，叶片褶皱、变红，小叶，顶芽异常激增，花变叶。已发现16SrⅠ组（Aster yellows group）、16SrⅡ组（Peanut witches'-broom group）和16SrⅤ组（Elm yellows group）植原体的侵染。

密蒙花

密蒙花（*Buddleja officinalis*）是马钱科（Loganiaceae）醉鱼草属的灌木，在我国广泛分布。密蒙花植原体病害仅在我国云南有发生报道。密蒙花感染植原体后，小叶、叶片黄化，大量细弱的枝条丛生成簇，矮化。已发现16SrⅫ组（Stolbur group）植原体的侵染。

墨苜蓿

墨苜蓿（*Richardia scabra*）是茜草科（Rubiaceae）墨苜蓿属的一年生直立草本，在我国广东、海南、香港等地广泛分布。墨苜蓿植原体病害仅在印度和我国的海南有发生报道。墨苜蓿感染植原体后，发育迟缓，小叶，花变叶，花绿变，花序畸形（图5-46）。已发现16SrⅡ组（Peanut witches'-broom group）和16SrⅥ组（Clover proliferation group）植原体的侵染。

图5-46　墨苜蓿植原体病害

牛筋草

牛筋草（*Eleusine indica*）又名蟋蟀草，是禾本科（Gramineae）䅟属的一年生直立草本，在我国南北方广泛分布。牛筋草植原体病害在韩国、印度、泰国、缅甸、肯尼亚及我国的台湾和山东有发生报道。牛筋草感染植原体后，叶片黄化，小叶，枝芽大量增加呈丛枝状。已发现16SrⅠ组（Aster yellows group）和16SrⅩⅣ组（Bermudagrass white leaf group）植原体的侵染。

青葙

青葙（*Celosia argentea*）又名狗尾草、百日红、鸡冠花，是苋科（Amaranthaceae）青葙属的一年生草本，在我国分布广泛。青葙植原体病害在印度、巴西及我国的海南和台湾有发生报道。青葙感染植原体后，枝芽异常增殖，扁茎，节间缩短（图5-47）。已发现16SrⅠ组（Aster yellows group）、16SrⅡ组（Peanut witches'-broom group）和16SrⅢ组（X-disease group）植原体的侵染。

图5-47　青葙植原体病害（陈旺提供）

赛葵

赛葵（*Malvastrum coromandelianum*）又名黄花棉、黄花草，是锦葵科（Malvaceae）

赛葵属的亚灌木状草本，在我国台湾、福建、广东、广西和云南等地广泛分布。赛葵植原体病害仅在我国海南有发生报道。赛葵感染植原体后，叶片变小，花变叶。已发现16SrⅠ组（Aster yellows group）植原体的侵染。

三点金

三点金（*Grona triflora*）又名蝇翅草、三点金草，是豆科（Fabaceae）假地豆属的多年生草本，广泛分布于热带及亚热带地区。三点金植原体病害仅在我国台湾有发生报道。三点金感染植原体后，植株表现丛枝、小叶、花变叶、花绿变等症状。已发现16SrⅡ组（Peanut witches'-broom group）植原体的侵染。

蛇婆子

蛇婆子（*Waltheria indica*）又名和他草，是锦葵科（Malvaceae）蛇婆子属略直立或匍匐状半灌木，主要分布于热带和亚热带地区。蛇婆子植原体病害仅在巴西和我国的海南有发生报道。蛇婆子感染植原体后，叶片褪绿，小叶，花绿变（图5-48）。已发现16SrⅠ组（Aster yellows group）植原体的侵染。

图5-48 蛇婆子植原体病害

菽麻

菽麻（*Crotalaria juncea*）是豆科（Leguminosae）猪屎豆属的直立草本，在我国福建、台湾、广东、广西等地有栽培。菽麻植原体病害在澳大利亚、印度、巴西和我国的云南有发生报道。菽麻感染植原体后，枝芽异常激增呈簇生状，花绿变。已发现16SrⅠ组（Aster yellows group）、16SrⅡ组（Peanut witches'-broom group）、16SrⅢ组（X-disease

野茼蒿

野茼蒿（*Crassocephalum crepidioides*）又名冬风菜、革命菜，是菊科（Asteraceae）野茼蒿属的直立草本，广泛分布于热带地区。野茼蒿感染植原体后，植株矮化，枝芽丛生，小叶，花序聚生，花器绿变（图5-53）。已发现16SrⅡ组（Peanut witches'-broom group）植原体的侵染。

图5-53　野茼蒿植原体病害

叶下珠

叶下珠（*Phyllanthus urinaria*）是叶下珠科（Phyllanthaceae）叶下珠属的一种杂草，在我国热区均有分布。叶下珠植原体病害仅在我国海南有发生报道。叶下珠感染植原体后，枝芽异常激增，小叶，丛枝（图5-54）。已发现16SrⅠ组（Aster yellows group）植原体的侵染。

图5-54　叶下珠植原体病害

赛葵属的亚灌木状草本，在我国台湾、福建、广东、广西和云南等地广泛分布。赛葵植原体病害仅在我国海南有发生报道。赛葵感染植原体后，叶片变小，花变叶。已发现16Sr Ⅰ组（Aster yellows group）植原体的侵染。

三点金

三点金（*Grona triflora*）又名蝇翅草、三点金草，是豆科（Fabaceae）假地豆属的多年生草本，广泛分布于热带及亚热带地区。三点金植原体病害仅在我国台湾有发生报道。三点金感染植原体后，植株表现丛枝、小叶、花变叶、花绿变等症状。已发现16Sr Ⅱ组（Peanut witches'-broom group）植原体的侵染。

蛇婆子

蛇婆子（*Waltheria indica*）又名和他草，是锦葵科（Malvaceae）蛇婆子属略直立或匍匐状半灌木，主要分布于热带和亚热带地区。蛇婆子植原体病害仅在巴西和我国的海南有发生报道。蛇婆子感染植原体后，叶片褪绿，小叶，花绿变（图5-48）。已发现16Sr Ⅰ组（Aster yellows group）植原体的侵染。

图5-48　蛇婆子植原体病害

菽麻

菽麻（*Crotalaria juncea*）是豆科（Leguminosae）猪屎豆属的直立草本，在我国福建、台湾、广东、广西等地有栽培。菽麻植原体病害在澳大利亚、印度、巴西和我国的云南有发生报道。菽麻感染植原体后，枝芽异常激增呈簇生状，花绿变。已发现16Sr Ⅰ组（Aster yellows group）、16Sr Ⅱ组（Peanut witches'-broom group）、16Sr Ⅲ组（X-disease group）、16Sr Ⅶ组（Ash yellows group）和16Srr Ⅸ组（Pigeon pea witches'-broom group）植原体的侵染。

甜麻

甜麻（*Corchorus aestuans*）又名假黄麻、针筒草，是锦葵科（Malvaceae）黄麻属的一年生草本，在我国长江以南各地广泛分布。甜麻植原体病害仅在我国海南有发生报道。甜麻感染植原体后，叶片变红，枝芽异常增殖呈丛簇状，花变叶（图5-49）。已发现16Sr Ⅱ组（Peanut witches'-broom group）植原体的侵染。

图5-49　甜麻植原体病害（陈旺提供）

细圆藤

细圆藤（*Pericampylus glaucus*）又名广藤，是防己科（Menispermaceae）细圆藤属的木质藤本，在我国长江流域以南广泛分布。细圆藤植原体病害仅在我国海南有发生报道。细圆藤感染植原体后，叶片褪绿，丛枝，节间变短（图5-50）。已发现16Sr Ⅰ组（Aster yellows group）植原体的侵染。

图5-50　细圆藤植原体病害

小驳骨

小驳骨（*Justicia gendarussa*）是爵床科（Acanthaceae）爵床属的多年生直立草本或亚灌木，主要分布在热带和亚热带地区。小驳骨植原体病害仅在印度和我国云南有发生报道。小驳骨感染植原体后，叶片小而黄，侧芽细弱，枝条丛生成簇状，节间明显缩短，植株严重矮化。已发现16SrⅡ组（Peanut witches'-broom group）和16SrⅩⅣ组（Bermudagrass white leaf group）植原体的侵染。

小心叶薯

小心叶薯（*Ipomoea obscura*）是旋花科（Convolvulaceae）番薯属的草本植物，主要分布在热带和亚热带地区。小心叶薯植原体病害仅在美国、我国的台湾和海南有发生报道。小心叶薯感染植原体后，叶片小，丛枝，节间变短（图5-51）。已发现16SrⅠ组（Aster yellows group）和16SrⅡ组（Peanut witches'-broom group）植原体的侵染。

图5-51　小心叶薯植原体病害

烟草

烟草是茄科（Solanaceae）烟草属（*Nicotiana*）植物的统称，一年生草本。烟草原产于南美洲，但在全世界都有商业种植。烟草植原体病害仅在墨西哥、印度、土耳其和我国台湾有发生报道。烟草感染植原体后，芽激增呈丛枝状，小叶，花绿变、花变叶，矮化（图5-52）。已发现16SrⅠ组（Aster yellows group）、16SrⅡ组（Peanut witches'-broom group）、16SrⅢ组（X-disease group）和16SrⅫ组（Stolbur group）植原体的侵染。

图5-52　烟草植原体病害（Liao et al.，2022）

野茼蒿

野茼蒿（*Crassocephalum crepidioides*）又名冬风菜、革命菜，是菊科（Asteraceae）野茼蒿属的直立草本，广泛分布于热带地区。野茼蒿感染植原体后，植株矮化，枝芽丛生，小叶，花序聚生，花器绿变（图5-53）。已发现16SrⅡ组（Peanut witches'-broom group）植原体的侵染。

图5-53　野茼蒿植原体病害

叶下珠

叶下珠（*Phyllanthus urinaria*）是叶下珠科（Phyllanthaceae）叶下珠属的一种杂草，在我国热区均有分布。叶下珠植原体病害仅在我国海南有发生报道。叶下珠感染植原体后，枝芽异常激增，小叶，丛枝（图5-54）。已发现16SrⅠ组（Aster yellows group）植原体的侵染。

图5-54　叶下珠植原体病害

夜香牛

夜香牛（*Cyanthillium cinereum*）又名染色草、伤寒草、寄色草，是菊科（Asteraceae）夜香牛属的一年生或多年生草本，在我国广东、广西、云南、台湾、福建等地广泛分布，常见于山坡旷野、荒地、田边、路旁。夜香牛感染植原体后，主要表现为丛枝症状（图5-55）。已发现16Sr Ⅰ组（Aster yellows group）植原体的侵染。

图5-55　夜香牛植原体病害

银胶菊

银胶菊（*Parthenium hysterophorus*）是菊科（Asteraceae）银胶菊属的一年生草本，在我国广东、广西、贵州、云南、海南等地广泛分布。银胶菊植原体病害在印度、越南、巴基斯坦和我国海南有发生报道。银胶菊感染植原体后，植株发育不良，枝芽激增，花序簇生，花瓣变绿，小叶和丛枝（图5-56）。已发现16Sr Ⅰ组（Aster yellows group）和16Sr Ⅱ组（Peanut witches'-broom group）植原体的侵染。

图5-56　银胶菊植原体病害

长序大豆

长序大豆（*Neonotonia wightii*）是豆科（Fabaceae）长序大豆属植物，在我国台湾、云南等地有分布。长序大豆植原体病害仅在我国云南有发生报道。长序大豆感染植原体后，植株发育不良，嫩枝过多，花序簇生，花绿变，小叶和丛枝。已发现16SrⅡ组（Peanut witches'-broom group）植原体的侵染。

长序苋

长序苋（*Digera muricata*）又名阿维尔长序苋，是苋科（Amaranthaceae）长序苋属植物，在我国台湾、安徽、河南等地有分布。长序苋植原体病害仅在印度和我国台湾有发生报道。长序苋感染植原体后，嫩芽丛簇，花绿变，花变叶（图5-57）。已发现16SrⅠ组（Aster yellows group）植原体的侵染。

图5-57　长序苋植原体病害（Mejia et al.，2022）

中华苦荬菜

中华苦荬菜（*Ixeris chinensis*）又名山鸭舌草、山苦荬、苦麻子，是菊科（Asteraceae）苦荬菜属的多年生草本，在我国分布广泛。中华苦荬菜植原体病害仅在我国的台湾和陕西有发生报道。植原体侵染中华苦荬菜后，枝芽异常增殖，花变叶，花绿变，紫顶，丛枝，扁枝。已发现16SrⅠ组（Aster yellows group）和16SrⅡ组（Peanut witches'-broom group）植原体的侵染。

皱子白花菜

皱子白花菜（*Cleome rutidosperma*）又名皱子鸟足菜，是白花菜科（Cleomaceae）鸟足菜属的一年生草本植物，在我国云南、台湾、广东、广西、海南等地广泛分布。皱子白花菜植原体病害仅在牙买加及我国的海南和台湾有发生报道。皱子白花菜感染植原体后，枝芽异常激增，小叶，丛枝，花绿变（图5-58）。已发现16SrⅡ组（Peanut witches'-broom group）和16SrⅣ组（Coconut lethal yellows group）植原体的侵染。

图5-58　皱子白花菜植原体病害（陈旺提供）

参考文献

蔡红，2007. 云南省植原体株系及其相关病害的多样性研究[D]. 昆明：云南农业大学.

蔡红，孔宝华，陈海如，2003. 长春花黄化植原体（PY）株系的检测与鉴定[J]. 微生物学报，43（1）：116-119.

蔡红，李凡，孔宝华，等，2001. 仙人掌丛枝病植原体CWB1株系16S rRNA基因克隆及序列分析[J]. 微生物学报，41（6）：693-698.

蔡红，李小林，孔宝华，等，2005a. 黄槐丛枝病植原体的检测及鉴定[J]. 植物病理学报，35（1）：19-23.

蔡红，张华明，陈海如，等，2005b. 竹丛枝植原体16S rDNA片段克隆与序列分析[J]. 植物保护，31（2）：38-40.

蔡红，张华明，吴华英，等，2005c. 一种引起香石竹黄化病植原体的初步鉴定[J]. 植物病理学报（S1）：151-152.

蔡银枫，2020. 喜树丛枝植原体在拟帕小绿叶蝉体内的侵染循回研究[D]. 昆明：云南农业大学.

车海彦，2010. 海南省植原体病害多样性调查及槟榔黄化病植原体的分子检测技术研究[D]. 杨凌：西北农林科技大学.

车海彦，罗大全，符瑞益，等，2009. 海南长春花黄化病植原体的16S rDNA序列分析研究[J]. 植物病理学报，39（2）：212-216.

车海彦，罗大全，符瑞益，等，2010. 海南省木豆丛枝病植原体的分子检测及鉴定[J]. 植物病理学报，40（2）：113-121.

车海彦，吴翠婷，符瑞益，等，2010. 海南槟榔黄化病病原物的分子鉴定[J]. 热带作物学报，31（1）：83-87.

车海彦，郑文虎，符瑞益，等，2011. 野茼蒿变叶病病原物的分子鉴定[J]. 热带作物学报，32（2）：289-292.

车海彦，郑文虎，温衍生，等，2011. 海南长春花变叶病病原物的分子鉴定[J]. 热带作物学报，32（3）：485-489.

陈健鑫，张宏雁，魏玉倩，等，2021. 滇朴丛芽病植原体的分子鉴定[J]. 植物保护，47（2）：70-77.

陈妮，2011. 山东省桑萎缩、枣疯病及其他三种植原体的分子检测与鉴定[D]. 泰安：山东农业大学.

段雅雯，杨毅，陆秋蕾，等，2019. 卵叶山蚂蝗丛枝植原体的分子鉴定[J]. 植物检疫，33（4）：29-32.

高三基，郭晋隆，陈如凯，等，2007. 福州地区甘蔗黄叶病病原分子鉴定及电镜检测[J]. 作物学报，33（7）：1210-1213.

何园歌，2016. 华南地区水稻橙叶病调查及其病原分子检测体系的建立[D]. 广州：华南农业大学.

和志娇，蔡红，陈海如，等，2005. 云南泡桐丛枝病植原体核糖体蛋白基因片段序列分析[J]. 植物病理学报（S1）：18-21.

兰平，李文凤，朱水芳，等，2001. 甘薯丛枝病植原体的PCR检测[J]. 植物学通报，18（2）：210-215.

李光明，袁坤，杨礼富，等，2012. 假臭草丛枝病植原体鉴定与多样性分析[J]. 生态学杂志，31（12）：3179-3186.

李横虹，邱并生，李锂，等，1999. 蕉束顶病植原体16S rDNA片段的RFLP和序列分析[J].

微生物学报，39（4）：315-320.

李龙，刘乐荷，伍建榕，等，2018. 车桑子丛枝病的病原分子检测鉴定及风险分析[J]. 中国森林病虫，37（3）：1-6.

李文凤，王晓燕，仓晓燕，等，2022. 耿马甘蔗白叶病植原体昆虫介体调查与检测[J]. 农学学报，12（5）：6-9.

李文凤，王晓燕，黄应昆，等，2014. 云南蔗区发现由植原体引起的检疫性病害甘蔗白叶病[J]. 植物病理学报，44（5）：556-560.

李永，田国忠，朴春根，等，2011. 灰叶丛枝病病原的分子鉴定及系统发育分析[J]. 林业科学研究，24（5）：549-553.

李永，田国忠，徐启聪，等，2009. 臭矢菜丛枝病植原体的分子鉴定研究[J]. 植物病理学报，39（4）：377-384.

李永，徐启聪，田国忠，等，2010. 猪屎豆丛枝病植原体的分子检测与鉴定[J]. 林业科学，46（1）：163-168.

李正刚，佘小漫，汤亚飞，等，2019. 广东枣疯病植原体的鉴定[J]. 植物病理学报，49（2）：281-282.

李正男，鞠明岫，马强，等，2020. 菊花绿变病植原体在中国的发生及分子鉴定[J]. 植物检疫，34（3）：37-42.

李正男，马强，张磊，等，2021. 宁夏马铃薯僵顶植原体的分子鉴定[J]. 植物保护，47（2）：56-61.

梁静思，祝菊澧，徐裴，等，2020. 植原体及马铃薯相关病害研究进展[J]. 园艺学报，47（9）：1777-1792.

林积秀，严叔平，陈红运，等，2013. 福建永安莴苣上一种新病害的研究简报[J]. 中国植保导刊（11）：33-34.

林积秀，严叔平，纪翠红，等，2016. 莴苣褪绿心腐病综合防控技术[J]. 长江蔬菜（11）：51-52.

林兆威，牛晓庆，唐庆华，等，2024a. 叶下珠黄化植原体和丛枝植原体的分子鉴定[J]. 热带作物学报，45（4）：864-871.

林兆威，牛晓庆，于少帅，等，2024b. 苦楝黄化植原体的分子鉴定[J]. 分子植物育种，22（1）：117-125.

刘建密，纪翠红，陈细红，等，2015. 福建永安市辣椒丛枝病植原体的分子检测及鉴定[J]. 植物保护，41（3）：115-118，153.

刘俊男，王柱华，万琼莲，等，2020. 萝卜花变叶病病原鉴定及赤霉素合成与代谢途径中关键酶基因的表达[J]. 云南农业大学学报：自然科学版，35（4）：614-621.

刘清神，许东林，林盘芳，等，2006. 广州桑树植原体分子检测及多样性初探[J]. 蚕业科学，32（1）：1-5.

刘涛，蔡红，赵丹，等，2007. 紫花苜蓿丛枝病植原体的分子检测及鉴定[J]. 植物病理学报，37（4）：362-367.

鹿鹏鹏，2021. 剑麻紫色卷叶病相关植原体的检测技术及其在植株体内变化规律的研究[D]. 海口：海南大学.

罗大全，车海彦，刘先宝，等，2008. 海南苦楝丛枝病植原体的分子鉴定[J]. 热带作物学报，29（4）：522-524.

罗大全，陈慕容，叶沙冰，等，2001. 海南槟榔黄化病的病原鉴定研究[J]. 热带作物学报，22（2）：43-46.

石宝萍，李成亮，都业娟，等，2013. 莴苣黄化病植原体的分子鉴定[J]. 石河子大学学报：自然科学版，31（6）：669-674.

时涛，刘先宝，黄贵修，2015. 木薯丛枝病和蛙皮病入侵我国的风险分析[J]. 热带生物学报，6（4）：432-437.

苏帆，杨子祥，王柱华，等，2021. 云南芝麻丛枝和花变叶植原体16S rRNA和 rp 基因序列分析[J]. 植物病理学报，51（6）：888-897.

汤亚飞，李正刚，佘小漫，等，2024. 广东番茄巨芽病植原体的分子鉴定[J]. 植物保护，50（1）：203-210.

汤亚飞，林祺，佘小漫，等，2022. 广东花生丛枝病植原体的分子鉴定[J]. 植物保护，48（5）：83-90.

唐伟文，骆学海，张曙光，等，1986. 海南岛番茄巨芽病和番茄丛枝病的初步鉴定[J]. 植物病理学报，16（2）：105-108.

田文杰，2021. 贵州樱桃植原体的分子检测技术[D]. 贵阳：贵州大学.

万琼莲，2013. 云南元谋植原体病害多位点基因分子特性及相关株系的保存研究[D]. 昆明：云南农业大学.

万琼莲，苏帆，许杏萍，等，2020. 一种山黄麻丛枝病的分子检测与鉴定[J]. 玉溪师范学院学报，36（3）：46-52.

万琼莲，王连春，王泉，等，2021. 羽脉山黄麻丛枝植原体的分子鉴定及病害调查[J]. 林业科学，57（5）：195-201.

万琼莲，杨子祥，王连春，等，2014. 云南花生丛枝植原体16S rRNA和核糖体蛋白基因序列分析[J]. 植物病理学报，44（4）：370-378.

王晓燕，张荣跃，李庆红，等，2023. 中国植原体病害的状况，分布及多样性研究进展[J]. 农学学报，13（3）：58-64.

王真辉，陈秋波，郭志立，等，2007. 假臭草丛枝病植原体16S rDNA检测与PCR-RFLP分析[J]. 热带作物学报，28（4）：51-56.

王柱华，刘俊男，杨子祥，等，2018. 云南蔓草虫豆丛枝植原体16S rDNA和*secY*基因序列分析[J]. 云南农业大学学报：自然科学版，33（5）：826-835.

王柱华，王文鹏，袁恩平，等，2021. 喜树丛枝植原体的分子鉴定及TaqMan探针实时荧光定量PCR检测方法的建立[J]. 植物病理学报，51（3）：429-440.

吴德喜，袁恩平，王连春，等，2013. 苦楝丛枝植原体相关膜蛋白基因分析及结构预测[J]. 云南农业大学学报：自然科学版，28（3）：310-316.

许杏萍，苏帆，杨子祥，等，2021. 云南小驳骨丛枝植原体的分子鉴定及相关基因序列分析[J]. 云南农业大学学报：自然科学版，36（2）：205-214.

杨海中，杨毅，陈剑山，2018. 赛葵花变叶植原体的鉴定及其16S rDNA序列特征分析[J]. 中国植保导刊，38（11）：21-24.

杨静宇，杨毅，车海彦，等，2015. 海南竹柏扁枝病病原植原体的分子检测鉴定[J]. 热带作物学报，36（6）：1147-1152.

杨毅，余希希，张力维，等，2021. 云南省马铃薯植原体发生特点及基因序列分析：植物病理科技创新与绿色防控——中国植物病理学会2021年学术年会论文集[C]. 北京：中国农业科学技术出版社.

尹跃艳，汪开化，徐忠志，等，2014. 云南芒果变叶病植原体的分子鉴定[J]. 植物病理学报，44（3）：332-336.

于少帅，李永，任争光，等，2017. 多位点序列分析揭示我国16SrⅠ组植原体不同株系间遗传变异和系统发育关系（英文）[J]. 林业科学，53（3）：105-118.

张驰，胡成慧，邱静思，等，2019. 基于高通量测序鉴定慈姑黄化病病原[J]. 南方农业学报，50（3）：578-584.

张春平，武占敏，李正男，等，2011. 竹小叶病植原体的分子鉴定[J]. 植物病理学报，41（1）：31-36.

张景宁，万汉彬，陈北光，等，1982. 桉树黄化病病原物的初步研究[J]. 华南农学院学报，（4）：45-47.

郑冠标，陈慕容，陈作义，等，1985. 橡胶树丛枝病研究初报[J]. 热带作物研究（1）：19-22.

郑文虎，2012. 海南四种植原体病害病原的分子鉴定及长春花相关病害植原体质粒、*sec*基因分析[D]. 海口：海南大学.

郑文虎，车海彦，杨毅，等，2014. 海南黄灯笼辣椒小叶病病原物的分子鉴定[J]. 热带作物学报，35（1）：142-146.

郑晓慧，朱国翱，王连春，等，2012. 樱桃花变绿病植原体的分子鉴定[J]. 植物病理学报，42（5）：546-550.

周国辉，许东林，2005. 广东桉树黄化（丛枝）病植原体分子鉴定与检测[J]. 植物保护学报，32（4）：387-391.

周国辉，许东林，蔡艳清，等. 三种植原体16S rRNA基因片段PCR扩增及序列分析[J]. 云南农业大学学报，18（4）：163-165.

周亚奎，甘炳春，张争，等，2010. 利用巢式PCR对海南槟榔（*Areca catechu* L.）黄化病的初步检测[J]. 中国农学通报，26（22）：381-384.

ABUHADEMA Y, SAYED E S T, ISMAIL M, et al., 2021. Pathological changes in sweet basil (*Ocimum basilicum* L.) caused by mixed infection of tobacco mosaic virus and phytoplasma[J]. Egyptian Journal of Botany, 61（3）：823-836.

ALVAREZ E, MEJÍA J F, LLANO G A, et al., 2009. Characterization of a phytoplasma associated with frogskin disease in cassava[J]. Plant Disease, 93（11）：1139-1145.

CAI H, WANG L C, YANG Z X, et al., 2016. Evidence for the role of an invasive weed in widespread occurrence of phytoplasma diseases in diverse vegetable crops: implications from lineage-specific molecular markers[J]. Crop Protection, 89：193-201.

CAI H, WANG L, MU W, et al., 2016. Multilocus genotyping of a 'Candidatus Phytoplasma aurantifolia' -related strain associated with cauliflower phyllody disease in China[J]. Annals of applied biology, 169（1）：64-74.

CAI H, WEI W, DAVIS R E, et al., 2008. Genetic diversity among phytoplasmas infecting Opuntia species: virtual RFLP analysis identifies new subgroups in the peanut witches'-broom phytoplasma group[J]. International Journal of Systematic and Evolutionary Microbiology, 58（6）：1448-1457.

CATAL M, IKTEN C, YOL E, et al., 2013. First report of a 16SrIX group (Pigeon Pea Witches'-Broom) phytoplasma associated with sesame phyllody in Turkey[J]. Plant Disease, 97（6）：835.

CHE H, LI Z, ZHANG L, et al., 2012. Detection and identification of 16SrII group phytoplasmas infecting *Stylosanthes* in China[J]. Journal of Phytopathology, 160（7-8）：437-439.

CHE H, YU S, CHEN W, et al., 2024. Molecular identification and characterization of novel taxonomic subgroups and new host plants in 16SrI and 16SrII group phytoplasmas and their evolutionary diversity on Hainan Island, China[J]. Plant Disease, 108（6）：1703-1718.

CHEN J, LV Z, WEI Y, et al., 2022. First Report of 'Candidatus Phytoplasma asteris' associated with witches'-Broom disease of *Macadamia ternifolia* in China[J]. Plant Disease, 106（1）：311.

CHEN J, PU X, DENG X, et al., 2009. A phytoplasma related to 'Candidatus Phytoplasma asteris' detected in citrus showing Huanglongbing (yellow shoot disease) symptoms in Guangdong, PR China[J]. Phytopathology, 99（3）：236-242.

CHEN J, WU F, WU Z, et al., 2023. First Report of 'Candidatus Phytoplasma asteris' associated with witches'-broom disease of Cinnamomum camphora in China[J]. Plant Disease, 107（6）: 1934.

CHEN J, WU Z, WU F, et al., 2023a. First Report of 'Candidatus Phytoplasma asteris'-related strains associated with witches'-broom and plexus bud disease of Cerasus serrula in China[J]. Plant Disease, 107（10）: 3274.

CHEN J, WU Z, YU Z, et al., 2023b. First Report of 'Candidatus Phytoplasma asteris' associated with witches'-broom disease of Pinus yunnanensis in China[J]. Plant Disease, 107（10）: 3276.

CHEN W Y, HUANG Y C, TSAI M L, et al., 2011. Detection and identification of a new phytoplasma associated with periwinkle leaf yellowing disease in Taiwan[J]. Australasian Plant Pathology, 40: 476-483.

CHEN W, LI Y, FANG X, 2020. Detection and molecular characterization of a phytoplasma in Eclipta prostrata in China[J]. Journal of General Plant Pathology, 86: 60-64.

CHEN W, LI Y, LIU F, et al., 2016. First report of a 16SrⅡ-a phytoplasma infecting Celosia argentea in China[J]. Journal of Plant Pathology, 98: 691.

CHEN W, LI Y, LIU F, et al., 2017. Melochia corchorifolia, a new host of 16SrⅠ-B phytoplasma in China[J]. Journal of Plant Pathology, 99: 291.

CHEN Y M, CHIEN Y Y, CHEN Y K, et al., 2021. Identification of 16SrⅡ-V phytoplasma associated with mungbean phyllody disease in Taiwan[J]. Plant Disease, 105（9）: 2290-2294.

CHENG M, DONG J, LASKI P J, et al., 2011. First report of clover proliferation group phytoplasmas（16SrⅥ-A）associated with purple top diseased potatoes（Solanum tuberosum）in China[J]. Plant Disease, 95（7）: 871.

CHIEN Y Y, TAN C M, KUNG Y C, et al., 2021a. Threeflower Tickclover（Desmodium triflorum）is a new host for peanut witches'-broom phytoplasma, a 16SrⅡ-V subgroup strain, in Taiwan[J]. Plant Disease, 105（1）: 209.

CHIEN Y Y, TAN C M, KUNG Y C, et al., 2021b. Ixeris chinensis is a new host for peanut witches'-broom phytoplasma, a 16SrⅡ-V subgroup strain, in Taiwan[J]. Plant Disease, 105（1）: 210.

CHIEN Y Y, TSAI M C, CHOU Y L, et al., 2020. Fringed spiderflower（Cleome rutidosperma）is a new host for purple coneflower witches'-broom phytoplasma, a 16SrⅡ-V subgroup strain in Taiwan[J]. Plant Disease, 104（4）: 1247.

CHIU Y C, LIAO P Q, MEJIA H M, et al., 2023. Detection, Identification and Molecular

Characterization of the 16SrⅡ-Ⅴ subgroup phytoplasma strain associated with *Pisum sativum* and *Parthenium hysterophorus* L. [J]. Plants, 12（4）: 891.

CHO S T, CHEN A P, CHOU S J, et al., 2023. Complete genome sequence of "*Candidatus* Phytoplasma cynodontis" GY2015, a plant pathogen associated with Bermuda grass white leaf disease in Taiwan[J]. Microbiology Resource Announcements, 12（10）: e00457-23.

DAVIS R I, HENDERSON J, JONES L M, et al., 2015. First record of a wilt disease of banana plants associated with phytoplasmas in Solomon Islands[J]. Australasian Plant Disease Notes, 10: 1-6.

DU Y J, MOU H Q, SHI B P, et al., 2013. Molecular detection and identification of a 16SrⅥ group phytoplasma associated with tomato big bud disease in Xinjiang, China[J]. Journal of Phytopathology, 161（11-12）: 870-873.

FENG Y C, HUNG T H, SU H J, 2015. Detection and inoculation of peanut witches'-broom phytoplasma（16SrⅡ-A）and periwinkle leaf yellowing phytoplasma（16SrⅠ-B）in citrus cultivars in Taiwan[J]. Journal of Phytopathology, 163（5）: 364-376.

GOH R P, LEE S, CHU C C, 2024. First report of a '*Candidatus* Phytoplasma australasiaticum'-related phytoplasma strain associated with shoot proliferation disease of variegated croton in Taiwan[J]. Plant Disease, 108（3）: 781.

KIRDAT K, TIWAREKAR B, THORAT V, et al., 2021. '*Candidatus* Phytoplasma sacchari', a novel taxon-associated with Sugarcane Grassy Shoot（SCGS）disease[J]. International Journal of Systematic and Evolutionary Microbiology, 71（1）: 004591.

LEE S, CHU C Y, CHU C C, 2021. Variability of phytoplasma infection density in poinsettia and evaluation of its association with the level of branching in host plants[J]. Plant Disease, 105（05）: 1539-1545.

LEE S, LEE Y J, CHANG C Y, et al., 2021. First report of a '*Candidatus* Phytoplasma aurantifolia'-related strain（16SrⅡ-Ⅴ）associated with phyllody, virescence, and shoot proliferation of sweet william（*Dianthus barbatus*）in Taiwan[J]. Plant Disease, 105（10）: 3285.

LI Y, CHEN W, 2018a. First report of a 16SrⅡ-A phytoplasma infecting *Spermacoce exilis* in China[J]. Journal of Plant Pathology, 100: 347.

LI Y, CHEN W, 2018b. Detection and identification of a '*Candidatus* Phytoplasma aurantifolia'-related strain（16SrⅡ-A subgroup）associated with *Corchorus aestuans* phyllody in China[J]. Journal of General Plant Pathology, 84: 243-245.

LI Z G, TANG Y F, SHE X M, et al., 2019. First report of 16SrⅡ-D phytoplasma associated with eggplant phyllody in China[J]. Canadian Journal of Plant Pathology, 41（3）: 339-

344.

LI Z, ZHANG L, CHE H, et al., 2011. A disease associated with phytoplasma in *Parthenium hysterophorus*[J]. Phytoparasitica, 39: 407-410.

LIAO P Q, CHEN Y K, MEJIA H M, et al., 2022. Detection, identification, and molecular characterization of a 16SrⅡ-Ⅴ subgroup phytoplasma associated with *Nicotiana plumbaginifolia*[J]. Plant Disease, 106（3）: 805-809.

LIAO P Q, CHIU Y C, MEJIA H M, et al., 2023. First report of 'Candidatus Phytoplasma aurantifolia' associated with the invasive weed *Eclipta prostrata* in Taiwan[J]. Plant Disease, 107（2）: 550.

LIN Z, SONG W, MENG X, et al., 2023. First report of 16SrⅠ-B subgroup related phytoplasma associated with yellows symptoms of *Rubus cochinchinensis* in China[J]. Plant Disease, 107（9）: 2838.

LIU C, HUANG H, HONG S, et al., 2015. Peanut witches'-broom（PnWB）phytoplasma-mediated leafy flower symptoms and abnormal vascular bundles development[J]. Plant Signaling and Behavior, 10（12）: e1107690.

LIU H L, CHEN C C, LIN C P, 2007. Detection and identification of the phytoplasma associated with pear decline in Taiwan[J]. European Journal of Plant Pathology, 117: 281-291.

LIU S L, LIU H L, CHANG S C, et al., 2011. Phytoplasmas of two 16S rDNA groups are associated with pear decline in Taiwan[J]. Botanical Studies, 52: 313-320.

LONG J Y, CHEN Y H, XIA J R, 2007. First report of a group 16SrⅠ phytoplasma associated with *Amaranthus hypochondriacus* Cladodes in China[J]. Plant Disease, 95（7）: 871.

MEJIA H M, LIAO P Q, CHEN Y K, et al., 2022. Detection, identification, and molecular characterization of the 16SrⅡ-Ⅴ subgroup phytoplasma strain associated with *Digera muricata* in Taiwan[J]. Plant Disease, 106（7）: 1788-1792.

MIYAZAKI A, SHIGAKI T, KOINUMA H, et al., 2018. 'Candidatus Phytoplasma noviguineense', a novel taxon associated with Bogia coconut syndrome and banana wilt disease on the island of New Guinea[J]. International Journal of Systematic and Evolutionary Microbiology, 68（1）: 170-175.

MOU H Q, XU X, WANG R R, et al., 2014. *Salix tetradenia* Hand-Mazz: a new natural plant host of 'Candidatus Phytoplasma'[J]. Forest Pathology, 44（1）: 56-61.

ONG S, JONSON G B, CALASSANZIO M, et al., 2021. Geographic distribution, genetic variability and biological properties of rice orange leaf phytoplasma in Southeast Asia[J].

Pathogens, 10（2）：169.

PONNAMMA K N, SOLOMON J J, RAJEEV M P, et al., 1997. Evidences for transmission of yellow leaf disease of arec palm, areca catechu L. by *Proutista moesta*（westwood）（Homoptera：derbidae）[J]. Journal of Plantation Crops, 25（2）：197-200.

RAO G P, KUMAR M, 2017. World status of phytoplasma diseases associated with eggplant[J]. Crop Protection, 96：22-29.

TEIXEIRA D C, WULFF N A, MARTINS E C, et al., 2009. A phytoplasma related to 'Candidatus Phytoplasma asteri' detected in citrus showing Huanglongbing（yellow shoot disease）symptoms in Guangdong, PR China[J]. Phytopathology, 99（3）：236-242.

TSENG Y W, CHANG C J, CHEN J W, et al., 2014. First report of a 16SrⅠ group phytoplasma associated with roselle（*Hibiscus sabdariffa*）wrinkled leaves and phyllody disorder in Taiwan[J]. Plant Disease, 98（7）：991.

TSENG Y W, CHANG C J, JAN F J, 2024. First report of 'Candidatus Phytoplasma australasiaticum' associated with phyllody, virescence, and witches'-broom disease in *Chrysanthemum morifolium* in Taiwan[J]. Plant Disease, 108（4）：1093.

TSENG Y W, CHANG H H, CHANG C J, et al., 2022. First Report of 'Candidatus Phytoplasma asteris'（16SrⅠ group）associated with *Murraya exotica* witches'-broom disease in Taiwan[J]. Plant Disease, 106（12）：3199.

TSENG Y W, DENG W L, CHANG C J, et al., 2012. First Report of a 16SrⅡ-A Subgroup phytoplasma associated with purple coneflower（*Echinacea purpurea*）witches'-broom disease in Taiwan[J]. Plant Disease, 96（4）：58.

TSENG Y W, DENG W L, CHANG C J, et al., 2016. The phytoplasma associated with purple woodnettle witches'-broom disease in Taiwan represents a new subgroup of the aster yellows phytoplasma group[J]. Annals of Applied Biology, 169（2）：298-310.

TSENG Y W, YEN J H, CHANG C J, et al., 2020. First report of a 'Candidatus Phytoplasma aurantifolia' strain associated with virescence, floral proliferation, and dwarf symptoms on Indigofera suffruticosa in Taiwan[J]. Plant Disease, 104（6）：1852-1852.

WANG B, LIN Z, ZHAO Z, et al., 2024. First Report of 16SrⅡ group-related phytoplasma associated with witches'-broom disease on cowpea（*Vigna unguiculata*）in Hainan province, China[J]. Plant Disease, 108（3）：783.

WANG C J, CHIEN Y Y, LIAO P Q, et al., 2021. First report of 16SrⅡ-Ⅴ phytoplasma associated with green manure soybean（*Glycine max*）in Taiwan[J]. Plant Disease, 105（7）：2012.

WANG G, WU W, XI J, et al., 2023. Detection and molecular identification of a 16SrⅠ group phytoplasma associated with sisal purple leafroll disease[J]. Plant Protection Science, 59（1）: 19-30.

WANG Q C, VALKONEN J P T, 2008. Efficient elimination of sweetpotato little leaf phytoplasma from sweet potato by cryotherapy of shoot tips[J]. Plant Pathology, 2008, 57（2）: 338-347.

WANG X, WANG C G, LI X Y, et al., 2021. Molecular detection and identification of a 'Candidatus Phytoplasma solani' -related strain associated with pumpkin witches'-broom in Xinjiang, China[J]. Phytopathologia Mediterranea, 2021, 60（1）: 63-68.

WEI W, ZHAO Y, 2022. Phytoplasma taxonomy: nomenclature, classification, and identification[J]. Biology, 2022, 11（8）: 1119.

WENG Y Y, LIOU W C, CHIEN Y Y, et al., 2021. First report of 16SrⅡ-Ⅴ peanut witches'-broom phytoplasma in snake gourd (Trichosanthes cucumerina) in Taiwan[J]. Plant Disease, 2021, 105（8）: 2236.

WU W, CAI H, WEI W, et al., 2012. Identification of two new phylogenetically distant phytoplasmas from Senna surattensis plants exhibiting stem fasciation and shoot proliferation symptoms[J]. Annals of Applied Biology, 160（1）: 25-34.

XI Y M, WANG S J, WANG S K, et al., 2022. First Report of phytoplasma from the 16SrⅤ Group associated with witches'-broom disease of Taxillus chinensis in China[J]. Plant Disease, 106（12）: 3199.

XU X, MOU H Q, ZHU S F, et al., 2013. Detection and characterization of phytoplasma associated with big bud disease of tomato in China[J]. Journal of Phytopathology, 161（6）: 430-433.

XUAN Z, CHE H Y, CAO X R, et al., 2019. First report of phytoplasma infecting Pterocarpus indicus in China[J]. Plant Disease, 103（4）: 759.

YAN X H, LIN J, LIU Y, et al., 2023. Complete genome sequence of "Candidatus Phytoplasma asteris" QS2022, a plant pathogen associated with lettuce chlorotic leaf rot disease in China[J]. Microbiology Resource Announcements, 12（6）: e00306-23.

YANG J Y, CHIEN Y Y, CHIU Y C, et al., 2023. Diversity, distribution, and status of phytoplasma diseases in Taiwan[M]//Diversity, Distribution, and Current Status. Academic Press: 149-168.

YANG Y, JIANG L, CHE H Y, et al., 2016a. Molecular identification of a 16SrⅡ-A group-related phytoplasma associated with cinnamon yellow leaf disease in China[J]. Journal of Phytopathology, 164（1）: 52-55.

YANG Y, JIANG L, CHE H, et al., 2016b. Identification of a novel subgroup 16SrⅡ-U phytoplasma associated with papaya little leaf disease[J]. International Journal of Systematic and Evolutionary Microbiology, 66（9）：3485-3491.

YANG Y, JIANG L, CHE H, et al., 2016c. Phytoplasma in association with rubber tree (*Hevea brasiliensis*) stem fasciation in China[J]. Plant Disease, 100（12）：2520.

YANG Y, JIANG L, TIAN Q, et al., 2017. Detection and identification of a novel subgroup 16SrⅡ-V phytoplasma associated with *Praxelis clematidea* phyllody disease[J]. International Journal of Systematic and Evolutionary Microbiology, 67（12）：5290-5295.

YU N T, XIE H M, WANG J H, et al., 2016. First report on the molecular identification of phytoplasma（16SrⅠ）associated with witches'-broom on *Dodonaea viscosa* in China[J]. Plant Disease, 100（6）：1232.

YU S S, SONG W W, 2024a. *Ipomoea obscura* represents a new host of phytoplasma belonging to 16SrⅡ group associated with witches'-broom symptoms in China[J]. Plant Disease, 108（3）：780.

YU S S, TANG Q H, FU D Q, et al., 2021a. Molecular identification and characterization of a phytoplasma strain associated with pepper (*Capsicum annuum* L.) yellow crinkle disease in China[J]. Journal of General Plant Pathology, 87（5）：330-334.

YU S S, TANG Q H, WU Y, et al., 2021b. First report of 16SrⅠ-B subgroup related phytoplasma associated with witches'-broom symptoms in *Pericampylus glaucus* in China[J]. Plant Disease, 105（2）：695.

YU S S, TANG Q H, WU Y, et al., 2021c. First report of 16SrXXXⅡ group related phytoplasma associated with *Trema tomentosa* witches'-broom disease in China[J]. Plant Disease, 105（4）：1191.

YU S S, WU Y, SONG W W, 2022. Occurrence of a 16SrⅡ-V subgroup phytoplasma associated with witches'-broom disease in *Melochia corchorifolia* in China[J]. Plant Disease, 106（2）：754.

YU S S, ZHAO R L, LIN M X, et al., 2021d. Occurrence of phytoplasma belonging to 16SrⅡ Group associated with witches'-broom symptoms in *Emilia sonchifolia* in Hainan Island of China[J]. Plant Disease, 105（12）：4151.

YU S S, ZHAO R L, LIN M X, et al., 2021e. *Tephrosia purpurea* represents a new host of 16SrⅡ-V subgroup phytoplasma associated with witches'-broom disease in China[J]. Plant Disease, 105（8）：2235.

YU S S, ZHAO R L, LIN M X, et al., 2021f. *Waltheria indica* is a new host of phytoplasma belongs to 16SrⅠ-B subgroup associated with virescence symptoms in China[J]. Plant Disease（7），105：2012.

YU S S, ZHU A N, SONG W W, et al., 2022. Molecular identification and characterization of two groups of phytoplasma and *Candidatus* Liberibacter asiaticus in single or mixed infection of *Citrus maxima* on Hainan Island of China[J]. Biology, 11（6）: 869.

YU S S, ZHU A N, SONG W W, 2023. *Carica papaya* represents a new host of 16SrⅠ-B subgroup phytoplasma associated with yellow symptoms in China[J]. Plant Disease, 107（1）: 211.

YU S S, ZHU A N, SONG W W, 2024. *Alocasia macrorrhiza* represents a new host of '*Candidatus* phytoplasma asteris'-related strains associated with yellows symptoms in China[J]. Plant Disease, 108（2）: 516.

第六章

植原体的复合侵染

不同病原复合侵染植物，会加重植物病情，给相关植物病害的诊断、监测、防控带来困难。植原体除了可以单独侵染植物外，植原体不同16Sr组的株系或植原体与病毒、韧皮部杆菌、螺原体等病原菌也可以复合侵染植物，增加了植原体病害诊断检测、防控管理的困难性与复杂性，因此，明确病害的病原种类及是否存在复合侵染现象，是相关病害监测防控的基础。

一、不同组植原体的复合侵染

不同组植原体可以复合侵染植物导致病害发生。如Lee等（2009）报道，在韩国表现丛枝、小叶、黄化症状的枣树可被16SrⅠ组和16SrⅤ组植原体复合侵染。Sun等（2013）在中国河南省也发现表现丛枝症状的枣树可被16SrⅠ组和16SrⅤ组植原体复合侵染。Lee等（2023）基于16S rRNA、*rp*、*tuf*和*secA*基因序列分析发现，韩国济州岛的山杜英（*Elaeocarpus sylvestris*）可同时被16SrⅠ和16SrXXXⅡ组植原体侵染。枣树和泡桐在我国分布广泛，复合侵染会导致16SrⅠ组和16SrⅤ组植原体在枣树与泡桐间通过叶蝉等刺吸式口器昆虫相互传播，从而增加了枣疯病与泡桐丛枝病的防控难度。

二、植原体与病毒复合侵染

自然条件下，植物可被植原体与病毒复合侵染，加重植物病情及损失（Kumar et al., 2016b; Venkataravanappa et al., 2018; Abirami et al., 2022; Mitra et al., 2022; Tiwari et al., 2022; Mall et al., 2023）。在印度，茄子可被16SrⅡ组植原体与黄瓜花叶病毒（Cucumber mosaic virus）复合侵染（Abirami et al., 2022），可被16SrⅥ组植原体与菜豆金色花叶病毒（Begomovirus）或马铃薯X病毒及马铃薯Y病毒（Potato virus X和Potato virus Y）复合侵染（Kumar et al., 2016b; Venkataravanappa et al., 2018）。甘蔗可

被16SrⅠ组或16SrⅪ组植原体与甘蔗黄叶病毒（sugarcane yellow leaf virus）复合侵染表现黄化症状（Arocha et al.，1999；El Sayed et al.，2016；Nithya et al.，2020）。番木瓜可被16SrⅡ组植原体与番木瓜环斑病毒（Papaya ringspot virus）复合侵染（Arocha et al.，2009）。香蕉可被16SrⅠ组植原体与香蕉束顶病毒（Banana bunchy top virus）、香蕉条纹MY病毒（Banana streak MY virus）复合侵染（Mitra et al.，2022）。植原体或病毒引起的植物病害症状非常相似，难以区分。因此，明确植物病害是由植原体和病毒等单一病原体引起的，还是由多种病原体的混合感染引起的，是开展相关植原体或病毒病有效监测、管理的前提。

三、植原体与柑橘黄化病菌复合侵染

植原体与韧皮部杆菌'*Ca.* Liberibacter'均是通过刺吸式口器昆虫传播、尚难分离培养的原核致病细菌（于少帅等，2016；杨毅等，2020；Doi et al.，1967；Liefting et al.，2008；Dickinson & Hodgetts，2013；Satta et al.，2016；Arec-Leal et al.，2020；Sumner-Kalkun et al.，2020；Bao et al.，2021；Quiroga et al.，2022）。韧皮部杆菌与柑橘黄龙病和马铃薯斑纹片病有关（杨毅等，2020；Liefting et al.，2008；Munyaneza et al.，2010；EPPO，2012；Satta et al.，2016；Arec-Leal et al.，2020；Sumner-Kalkun et al.，2020；Bao et al.，2021；Quiroga et al.，2022）。Yu等（2022）基于16S rRNA和β-operon基因片段，发现中国海南柚子（*Citrus maxima*）可被16SrⅡ-Ⅴ组植原体与柑橘黄化病菌'*Ca.* Liberibacter asiaticus'复合侵染表现类似黄化病症状（图6-1）。16SrⅡ-Ⅴ组植原体与柑橘黄化病菌'*Ca.* Liberibacter asiaticus'在伊朗可复合侵染甜橙（*C. sinensis*）（Alizadeh et al.，2017）。表现黄化病症状的柑橘类作物，可能是由柑橘黄化病菌单独侵染引起的，也可能是由柑橘黄龙病菌与植原体复合侵染引起的（Teixeira et al.，2008；Chen et al.，2009；Arratia-Castro et al.，2014；Alizadeh et al.，2017）。柑橘黄化病菌'*Ca.* Liberibacter asiaticus'除了可以与16SrⅡ组植原体复合侵染柑橘类作物外，也可以与其他不同16Sr组的植原体复合侵染柑橘类作物（Teixeira et al.，2008；Chen et al.，2009；Luis-Pantoja et al.，2021；Arratia-Castro et al.，2014）。如在中国广东（Chen et al.，2009）和墨西哥（Arratia-Castro et al.，2014），表现黄化病症状的柑橘可被16SrⅠ组植原体与柑橘黄化病菌'*Ca.* Liberibacter asiaticus'复合侵染。在巴西，表现黄化病症状的柑橘可被16SrⅨ组植原体与柑橘黄化病菌'*Ca.* Liberibacter asiaticus'复合侵染（Teixeira et al.，2008）。植原体与柑橘黄化病菌引起的症状十分相似，因此，不能仅通过症状对病原种类进行判别，应结合分子技术判断相关症状是由单一病原引起还是多个病原复合侵染引起（Teixeira et al.，2008；Chen et al.，2009；Arratia-Castro et al.，2014；Luis-Pantoja et al.，2021；Yu et al.，2022a）。

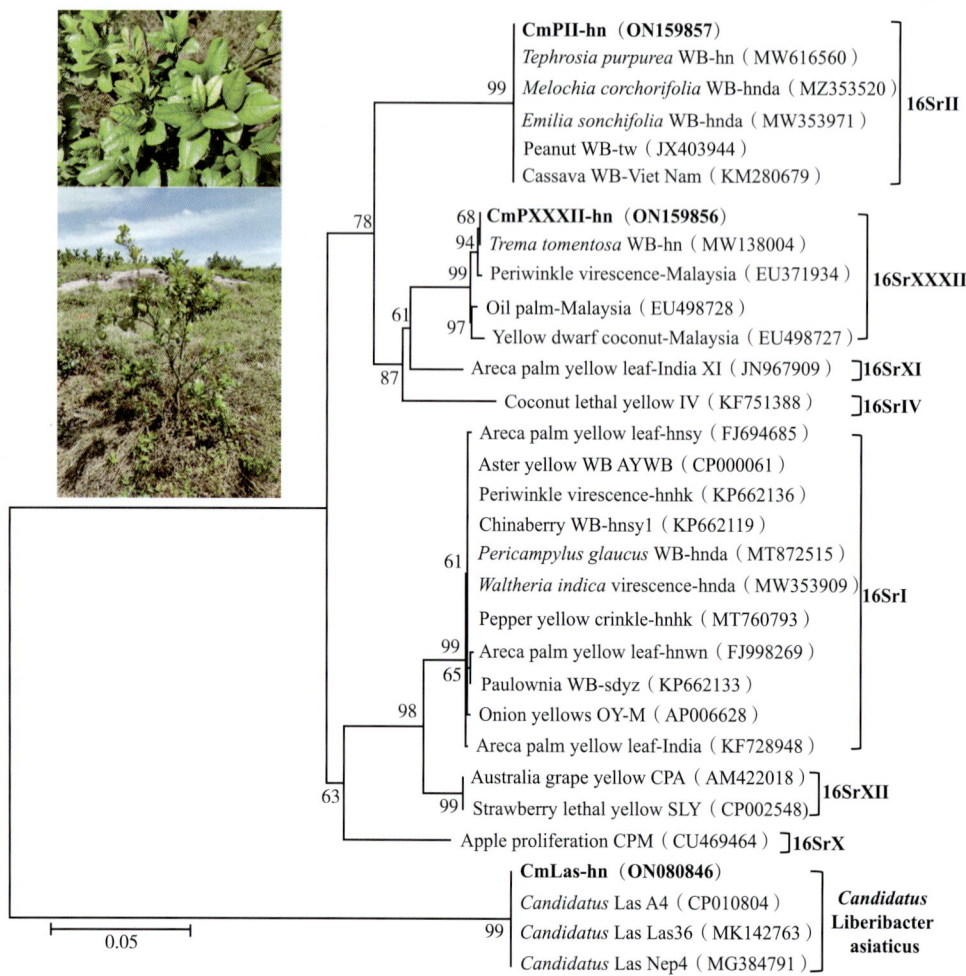

图6-1　海南植原体与韧皮部杆菌'*Ca*. Liberibacter asiaticus'复合侵染现象（Yu et al.，2022）

注：采用N-J法构建进化树，枝长代表遗传距离，自展值（1 000个重复）在分枝处表示。

四、植原体与马铃薯斑纹片病菌复合侵染

马铃薯斑纹片病菌'*Ca*. Liberibacter solanacearum'也是一种寄生于植物韧皮部、尚难分离培养的原核致病菌，可引起茄科和伞形科作物病害（Satta et al.，2016；Sumner-Kalkun et al.，2020）。该病原首次在新西兰的马铃薯上被发现（Liefting et al.，2008），随后在欧洲的胡萝卜和芹菜上也发现了该病原（Munyaneza et al.，2010；Alfaro-Fernández et al.，2012；EPPO，2012；Satta et al.，2016；Sumner-Kalkun et al.，2020）。Satta等（2016）基于16S rRNA基因分析发现，16SrⅠ组植原体可与马铃薯斑片病菌'*Ca*. Liberibacter solanacearum'复合侵染胡萝卜引起黄化症状。

五、植原体与螺原体复合侵染

螺原体（Spiroplasma）是一类基本形态为螺旋形、无细胞壁、能独立生活和自我复制的简单原核微生物（于汉寿等，2009；Davis et al.，1973；Schmitt et al.，1984；Harne et al.，2020；Galvão et al.，2021）。螺原体可以单独或与植原体复合侵染引起植物病害（Oliveira et al.，2015；Lebsky et al.，2019；Galvão et al.，2021；Salehi et al.，2022）。Lebsky等（2019）发现16SrⅣ组植原体可与螺原体复合侵染棕榈作物。Galvão等（2021）发现，在巴西，植原体可与螺原体复合侵染玉米导致玉米发育迟缓。在伊朗，16SrⅡ-C、16SrⅡ-D、16SrⅥ-A和16SrⅨ-C亚组植原体可与螺原体*Spiroplasma citri*复合侵染芝麻引起花变叶症状（Salehi et al.，2022）。

六、植原体与其他难培养病原菌复合侵染

植原体除了可以与病毒、韧皮部杆菌、螺原体等病原菌复合侵染外，也可与其他难培养微生物如'*Ca.* Phloemobacter fragariae'和'*Ca.* Arsenophonus phytopathogenicus'等复合侵染植物，加重植物病情（Gatineau et al.，2002；Danet et al.，2003；Behrmann et al.，2023；Duduk et al.，2023）。甜菜"basses richesses"综合症可由植原体与'*Ca.* Phloemobacter fragariae'复合侵染引起（Gatineau et al.，2002）。在德国，植原体'*Ca.* Phytoplasma solani'和难培养病原'*Ca.* Arsenophonus phytopathogenicus'可复合侵染甜菜和马铃薯（Behrmann et al.，2023；Duduk et al.，2023）。尽管由植原体、'*Ca.* Phloemobacter fragariae'或其他病原菌引起的植物病害的症状十分相似，但不同病原菌对植物的致病作用不同。如植原体与植物细胞坏死和细胞壁木质化有关，而'*Ca.* Phloemobacter fragariae'则与韧皮部细胞管腔中酚类化合物的沉积有关（Gatineau et al.，2002）。

七、植原体与真菌复合侵染

植原体不仅可以与原核生物一起侵染植物，也可以与真核生物如真菌等复合侵染植原体。Duduk等（2023）报道甜菜可被植原体'*Ca.* Phytoplasma solani'与病原真菌*Macrophomina phaseolina*复合侵染，植物病原真菌*Macrophomina phaseolina*与植原体复合侵染加剧了相关经济作物的损失。由此可知，植原体不同16Sr组的株系或植原体与其他不同原核和真核病原菌可以复合侵染植物引起相同或相似的植物病害。复合侵染所涉及病原的多样性也预示着在相关病害的监测与防控过程中需要多种方法系统、综合应用。

参考文献

杨毅，李丹阳，段雅雯，等，2020. 海南柑橘黄龙病发生分布调查及病原种类鉴定[J]. 植物检疫，34（3）：43-47.

于汉寿，刘淑园，阮康勤，等，2009. 螺原体的分类及其生物多样性研究进展[J]. 微生物学报，49（5）：567-572.

于少帅，徐启聪，林彩丽，等，2016. 植原体遗传多样性研究现状与展望[J]. 生物多样性，24（2）：205-215.

ABIRAMI R, MANORANJITHAM S K, MOHANKUMAR S, et al., 2022. Preponderance of mixed infection of *Cucumber mosaic virus* and 'Candidatus Phytoplasma australasia' on brinjal in India[J]. Microbial Pathogenesis, 169：105596.

ALFARO-FERNÁNDEZ A, CEBRIÁN M C, VILLAESCUSA F J, et al., 2012. First report of 'Candidatus Liberibacter solanacearum' in carrot in mainland Spain[J]. Plant Disease, 96：582.

ALIZADEH H, QUAGLINO F, AZADVAR M, et al., 2017. First report of a new citrus decline disease（CDD）in association with double and single infection by 'Candidatus Liberibacter asiaticus' and 'Candidatus Phytoplasma aurantifolia' related strains in Iran[J]. Plant Disease, 101：2145.

ARCE-LEAL A P, BAUTISTA R, RODRIGUEZ-NEGRETE E A, et al., 2020. Gene expression profile of Mexican Lime（*Citrus aurantifolia*）trees in response to Huanglongbing disease caused by *Candidatus* Liberibacter asiaticus[J]. Microorganisms, 8：528.

AROCHA Y, GONZALEZ L, PERALTA E L, et al., 1999. First report of virus and phytoplasma pathogens associated with yellow leaf syndrome of sugarcane in Cuba[J]. Plant Disease, 83：1177.

AROCHA Y, PIÑOL B, ACOSTA K, et al., 2009. Detection of phytoplasma and potyvirus pathogens in papaya（*Carica papaya* L.）affected with bunchy top symptom（BTS）in eastern Cuba[J]. Crop Protection, 28：640-646.

ARRATIA-CASTRO A A, SANTOS-CERVANTES M E, FERNÁNDEZ-HERRERA E, et al., 2014. Occurrence of 'Candidatus Phytoplasma asteris' in citrus showing Huanglongbing symptoms in Mexico[J]. Crop Protection, 62：144-151.

BAO M, ZHENG Z, CHEN J, et al., 2021. Investigation of citrus HLB symptom variations associated with "Candidatus Liberibacter asiaticus" strains harboring different phages in Southern China[J]. Agronomy, 11：2262.

BEHRMANN S C, RINKLEF A, LANG C, et al., 2023. Potato（*Solanum tuberosum*）as a new host for *Pentastiridius leporinus*（Hemiptera：Cixiidae）and *Candidatus* Arsenophonus Phytopathogenicus[J]. Insects, 14: 281.

CHEN J, PU X, DENG X, et al., 2009. A phytoplasma related to 'Candidatus Phytoplasma asteri' detected in citrus showing Huanglongbing（yellow shoot disease）symptoms in Guangdong, P. R. China[J]. Phytopathology, 99: 236-242.

DANET J L, FOISSAC X, ZREIK L, et al., 2003. 'Candidatus Phlomobacter fragariae' is the prevalent agent of marginal chlorosis of strawberry in French production fields and is transmitted by the planthopper *Cixius wagneri*（China）[J]. Phytopathology, 93: 644-649.

DAVIS R E, WORLEY J F, 1973. Spiroplasma: motile, helical microorganism associated with corn stunt disease[J]. Phytopathology, 63: 403-408.

DICKINSON M, HODGETTS J, 2013. Phytoplasma: Methods and Protocols[M]. Totowa, NJ, USA: Humana Press.

DOI Y, TERANAKA M, YORA K, et al., 1967. Mycoplasma-or PLT group-like microorganisms found in the phloem elements of plants infected with mulberry dwarf, potato witches'-broom, aster yellows, or paulownia witches'-broom[J]. Japanese Journal of Phytopathology, 33: 259-266.

DUDUK B, KOSOVAC A, STEPANOVIC J, et al., 2023. Phytoplasma, proteobacterium and fungus in single and mixed infections of sugar beet in central Europe[J]. Phytopathogenic Mollicutes, 13: 97-98.

EL SAYED A I, SOUFI Z, WAHDAN K M, et al. Detection and characterization of phytoplasma and sugarcane yellow leaf virus associated with leaf yellowing of sugarcane[J]. Journal of Phytopathology, 164: 217-225.

EPPO, 2012. First report of 'Candidatus Liberibacter solanacearum' on carrots and celery in Spain, in association with *Bactericera trigonica*[J]. EPPO Reporting Service-Pests and Diseases, 6: 4-5.

GALVÃO S R, SABATO E O, BEDENDO I P, 2021. Occurrence and distribution of single or mixed infection of phytoplasma and spiroplasma causing corn stunting in Brazil[J]. Tropical Plant Pathology, 46: 152-155.

GATINEAU F, JACOB N, VAUTRIN S, et al., 2002. Association with the syndrome "basses richesses" of sugar beet of a phytoplasma and a bacterium-like organism transmitted by a *Pentastiridius* sp[J]. Phytopathology, 92: 384-392.

HARNE S, GAYATHRI P, BÉVEN L, 2020. Exploring spiroplasma biology: opportunities and challenges[J]. Frontiers in Microbiology, 11: 589279.

KUMAR M, KATIYAR A, MADHUPRIYA, et al., 2016. First report of association of Potato virus X and Potato virus Y and 'Candidatus Phytoplasma trifolii' in brinjal in India[J]. Virus Disease, 27: 207-208.

LEBSKY V, HERNANDEZ J, BARRAZA A, et al., 2019. Ultrastructural analysis of spiroplasmas detected in palm species infected with the lethal yellowing phytoplasma from Yucatan and Baja California Sur, Mexico[J]. Phytopathogenic Mollicutes, 9: 155-156.

LEE G W, HAN S S, 2023. Molecular detection of phytoplasmas of the 16Sr I and 16Sr XXXII groups in *Elaeocarpus sylvestris* trees with decline disease in Jeju Island, South Korea[J]. The Plant Pathology Journal, 39: 149-157.

LEE S, HAN S, CHA B, 2009. Mixed infection of 16S rDNA I and V groups of phytoplasma in a single jujube tree[J]. The Plant Pathology Journal, 25: 21-25.

LIEFTING L W, EGUSQUIZA Z C, CLOVER G R G, et al., 2008. A new 'Candidatus Liberibacter' species in *Solanum tuberosum* in New Zealand[J]. Plant Disease, 92: 1474.

LUIS-PANTOJA M, PAREDES-TOMÁS C, UNEAU Y, et al., 2021. Identification of 'Candidatus Phytoplasma' species in "huanglongbing" infected citrus orchards in the Caribbean[J]. European Journal of Plant Pathology, 160: 185-198.

MALL S, VISHWAKARMA R, 2023. An update status of coinfection of phytoplasmas with other pathogens in plants[J]. Phytopathogenic Mollicutes, 13: 151-162.

MITRA S, DEBNATH P, RAI R, et al., 2022. Identification of 'Ca. Phytoplasma asteris', *Banana bunchy top virus* and *Banana streak MY virus* associated with Champa and Sabri banana cultivars in Tripura, a north eastern state of India[J]. European Journal of Plant Pathology, 163: 907-922.

MUNYANEZA J E, FISHER T W, SENGODA V G, et al., 2010. First report of 'Candidatus Liberibacter solanacearum' in carrots in Europe[J]. Plant Disease, 94: 639.

NITHYA K, PARAMESWARI B, VISWANATHAN R, 2020. Mixed infection of sugarcane yellow leaf virus and grassy shoot phytoplasma in yellow leaf affected Indian sugarcane cultivars[J]. The Plant Pathology Journal, 36: 364-377.

OLIVEIRA E, LANDAU E C, SOUSA S M, 2015. Simultaneous transmission of phytoplasma and spiroplasma by *Dalbulus maidis* leafhopper and symptoms of infected maize[J]. Phytopathogenic Mollicutes, 5: S99.

QUIROGA N, GAMBOA C, MEDINA G, et al., 2022. Survey for 'Candidatus Liberibacter' and 'Candidatus Phytoplasma' in *Citrus* in Chile[J]. Pathogens, 11: 48.

SALEHI M, FAGHIHI M M, SALEHI E, et al., 2022. Occurrence of single and mixed infection of *Spiroplasma citri* and phytoplasmas in sesame plants in Iran[J]. Australasian Plant Pathology, 51: 13-26.

SATTA E, RAMIREZ A S, PALTRINIERI S, et al., 2016. Simultaneous detection of mixed 'Candidatus Phytoplasma asteris' and 'Ca. Liberibacter solanacearum' infection in carrot[J]. Phytopathologia Mediterranea, 55: 401-409.

SCHMITT U, PETZOLD H, MARWITZ R, 1984. An in situ freeze-fracture study of Spiroplasma citri and the corn stunt Spiroplasma[J]. Phytopathology, 111: 297-304.

SUMNER-KALKUN J C, HIGHET F, ARNSDORF Y M, et al., 2020. 'Candidatus Liberibacter solanacearum' distribution and diversity in Scotland and the characterisation of novel haplotypes from Craspedolepta spp. (Psylloidea: Aphalaridae)[J]. Scientific Reports, 10: 16567.

SUN X C, MOU H Q, LI T T, et al., 2013. Mixed infection of two groups (16Sr I & V) of phytoplasmas in a single jujube tree in China[J]. Journal of Phytopathology, 161: 661-665.

TEIXEIRA D D C, WULFF N A, MARTINS E, et al., 2008. A phytoplasma closely related to the pigeon pea witches'-broom phytoplasma (16SrIX) is associated with citrus Huanglongbing symptoms in the State of São Paulo State, Brazil[J]. Phytopathology, 98: 977-984.

TIWARI N N, JAIN R K, PRAJAPATI M R, et al., 2022. Evidence of mixed infection of phytoplasma and Begomovirus associated with Withania somnifera and Capsicum annum plants from Uttar Pradesh, India[J]. Archives of Phytopathology and Plant Protection, 55: 2146-2157.

VENKATARAVANAPPA V, PRASANNA H C, LAKSHMINARAYANA C N, et al., 2018. Molecular detection and characterization of phytoplasma in association with Begomovirus in eggplant[J]. Acta Virologica, 62: 246-258.

YU S S, ZHU A N, SONG W W, et al., 2022. Molecular identification and characterization of two groups of phytoplasma and Candidatus Liberibacter asiaticus in single or mixed infection of Citrus maxima on Hainan Island of China[J]. Biology, 11: 869.

第七章 植原体病害防控管理

植原体病害侵染性强，寄主范围广泛，因而针对植原体病害的防治，应该贯彻"预防为主，综合防治"的植保方针，坚持统筹协调、因地制宜、分区治理、综合防控的原则。以农业防治为基础，协调运用物理、化学、生物、生态等各项防治措施对植原体病害进行科学、安全、有效的防控。

一、监测预警

植物病害监测预警是植保工作的重要组成部分，是制定病害防治措施的前提和基础，是实现作物病害科学精准防控，减少农药使用量的重要技术保障。传统的植原体病害监测方法主要通过田间人工调查获取监测数据。近年来，随着科学技术的发展，科研人员根据不同病害的发生特点，基于"3S"技术（遥感技术、地理信息系统、全球定位系统）、空气动力学技术、分子生物学技术、人工智能技术建立了多种植物病害监测预警系统，极大地促进了植物病害监测预警的准确度（胡小平等，2022；孙圆龙，2022；曹学仁等，2016）。如在槟榔黄化病防治方面，科研人员不仅集成了以症状识别和分子检测相结合的病害疫情监测技术体系，构建了病害疫情信息监测与共享平台（罗大全等，2001；罗大全等，2016），还利用遥感监测技术对槟榔黄化病进行监测（赵晋陵等，2020；金玉等，2020），这些监测预警技术的使用极大地提高了人们防治槟榔黄化病的水平。

二、农业防控

加强栽培管理、增加营养条件以及改进土壤水肥，可以促进植株根系发育，提高植株长势，增强植株的控害能力，对植原体病害防控具有明显的效果。如在桑树萎缩病和槟榔黄化病的发病初期，如果改进营养和水肥条件，植株的病害症状可以减轻，发病进程减缓（刘仲健等，1999；车海彦等，2018）。

种植无毒种苗可以减少田间初侵染源，从源头控制植原体病害的发生。如泡桐主要通过种根苗繁殖，一些无症带毒种根苗成为泡桐丛枝病传播的主要介体，种植无症带毒种根苗会导致幼龄树发病率上升，从而造成严重的经济损失。培育无毒苗木，大力推广脱毒苗育苗技术，便于有效控制植原体病害传播（田国忠等，2009；李艳玲等，2023）。

许多越年生和多年生杂草生育期长，往往成为植原体的中间寄主，也会成为许多传毒昆虫生长和繁殖的场所。清除苗床周边杂草，也是控制和预防植物植原体的一个重要农业措施。

种植抗病品种是防控植原体病害最为安全经济有效的措施（张曙光等，1999）。培育抗病品种一直是植原体病害防控工作的重点研究方向。如在抗枣疯病材料研究中，科研人员筛选出壶瓶枣、蛤蟆枣、婆婆枣、洪赵小枣、骏枣、清徐圆枣、秤砣枣和南京木枣单系等抗病种质，一些抗病种质已被应用于生产实践（赵锦等，2006；肖京等，2013；田国忠等，2013）。

三、物理防控

防虫网覆盖栽培是一项有效减缓植原体病害传播与蔓延的措施。在澳大利亚，Elder等（2002）和Walsh等（2006）通过在番木瓜种植地使用防虫网覆盖技术，有效控制了番木瓜植原体病害的发生蔓延。在母本园或者种苗生产园中使用防虫网，可以保证植株不受传毒昆虫的影响。但该方式的缺点较为明显，首先是隔离网很容易被木本植物和高大草本植物撕裂，其次在所有的热带地区，灼热的阳光辐射会使隔离网加速氧化并快速破损。

清除带毒植株或患病枝条是控制植原体病害的有效防控手段。彻底而及时地铲除枣疯树，尽量保持在生长季节里田间无绿色疯树叶的存在，结合其他枣树病虫的除治，兼治传病叶蝉，如此坚持数年后，即可在大流行的病区内将枣疯病年发病率压低至0.3%以下（刘仲健等，1999）。2009—2010年，海南省三亚市通过清除感病植株，一度控制了槟榔黄化病在三亚的蔓延速度。万宁市东兴农场一个种植面积超过200亩的槟榔园，从2013年起园主通过及时清除田间发病植株，有效控制了黄化病在槟榔园内的蔓延，保障了槟榔园的收益（车海彦等，2018）。山东省果树研究所对泡桐丛枝病进行病枝的环剥和修除试验，如果季节适当，治愈率一般可稳定在90%左右（刘仲健等，1999）。

四、化学防控

由于植原体是一类无细胞壁的原核生物，影响原核生物蛋白质或核酸合成的化合物一般都能抑制植原体的生长和繁殖，如四环素及四环素族抗生素，一般通过喷洒法、根部或枝条浸渍法及树干注射法施用。1968—1973年，日本一科研小组以桑树萎缩、水稻黄矮、马铃薯丛枝、翠菊黄化、莴苣黄化丛枝、檀香簇生等植原体病害为研究对象，开展四环素及其衍生物对植原体病害的治疗效果研究。结果表明，四环素及四环素衍生物能使植物植原体病害暂时减轻或缓解，但不能根治；如果植株呈现系统性症状，特别是老病株，四环素族的疗效也较差；植株根部较树冠部分易吸收抗生素，如果以溶液培养苗木，反复用药物浸根，则疗效最为显著；四环素族药物之间的防病效果差异不大；病株症状在用药后仅临时减轻或消失，即使在继续用药的情况下也常会重现（刘仲健等，1999）。罗大全等（2001）发现将土霉素和盐酸四环素注射到感染黄化病的槟榔茎秆中可有效抑制病情的发展。四环素、土霉素、螺旋霉素和红霉素等对枣疯病、泡桐丛枝病、桑萎缩病等都有明显的抑制作用。除四环素及四环素族抗生素外，链霉素、氯霉素、卡那霉素、新霉素等也曾被用于植原体病害的治疗试验，但效果大多不及四环素族抗生素。

除抗生素外，人们发现有些化合物对植原体病害也有一定的治疗效果。20世纪40年代，Stoddard对梨树X病进行醌氢醌、8-羟基喹啉硫酸盐和氢醌溶液浸渍治疗后，发芽嫩基外观健康，表现症状的苗木数量减少。一些磺胺类药物，如磺胺嘧啶、氨基苯磺酰胺、硫代异恶唑等可有效减轻翠菊黄化病症状，有效抑制甘蔗丛梢病的病情。磺胺类药物对植原体病害的疗效尚有待进一步的研究，阿的平、奎宁等药物对植原体也有一定的疗效（刘仲健等，1999）。

大多数植原体可通过叶蝉类、木虱、螨类等昆虫进行传播，因此积极采取防虫治虫措施是预防植物植原体病害发生和流行的重要手段之一。防虫治虫的前提是必须对传毒昆虫的种类、生活习性、传毒特点、迁飞规律等有全面和深入的了解，只有这样才能有的放矢，在适当的时候喷施吡虫啉、阿维菌素、溴氰菊酯、烯啶虫胺、噻虫嗪、啶虫脒等有效的杀虫剂，以杀灭媒介昆虫（陈怡光，2015）。

五、生物防控

在田间，许多植原体的传毒昆虫（叶蝉、飞虱等）会受到多种天敌的攻击，如蜘蛛、螨、蜂类、草蛉、瓢虫、步甲等，这些天敌对其具有很好的控制作用。蜘蛛是叶蝉和飞虱成虫和若虫的重要捕食者，而半翅目的盲蝽科则可能是卵的主要取食者，螯蜂科昆虫可产卵到叶蝉和飞虱的成虫和若虫体内，幼虫囊状物在寄主体外发育，其蛹则可掉入土内或落叶中发育。膜翅目的缨小蜂非常常见，但却很少看到它们攻击叶蝉和飞虱的卵，但其整个发育均在寄主卵内完成。双翅目头蝇科可专性寄生头喙亚目，它们可产卵于成虫和若虫体内，其幼虫在寄主体内发育，发育的后期可导致寄主腹部膨大。通过保护和利用这些天敌，可以有效地控制传毒昆虫的种群数量，进而控制植原体蔓延速度（Weintraub et al. 2010）。在湿度高的地区或季节，也可以通过在作物上喷洒含有白僵菌的生物制剂，从而降低叶蝉的虫口数量。

六、生态防控

大面积种植单一作物极易导致病害大流行，不同类型的作物间作或套种，可以增强田间生物多样性，改善农田生态环境，减少病虫发生。如槟榔园内合理的间作不仅能提高土地的利用率，抑制杂草的生长，改善土壤性状和园地生态环境，还可作为病虫害生态治理的一种手段，控制病虫害发生，减少化学农药的投入。在幼龄槟榔园内，若土壤肥力低，可间作柱花草、猪屎豆等绿肥植物来改善土壤性状，增加肥力。若土壤肥力高，可间作蔬菜、花生等短期经济作物；成龄槟榔园要充分考虑到株行距、光照等因素，可间作益智、胡椒、香草兰等经济作物，建立林下间作的生态防控模式，提高槟榔种植系统经济效益。建立林下及林缘间套种生态防控模式，同时在有条件的地方推广槟榔种养模式，提高槟榔种植系统的经济效益（车海彦等，2018）。

参考文献

曹学仁，周益林，2016. 植物病害监测预警新技术研究进展[J]. 植物保护，42（3）：1-7.

车海彦，曹学仁，禤哲，等，2018. 槟榔黄化病"该防"还是"该治"[J]. 中国热带农业，84（05）：48-50.

陈怡光，2015. 广东省罗定市水稻橙叶病发生原因分析及其防控对策[J]. 安徽农业科学（19）：87-88.

胡小平，户雪敏，马丽杰，等，2022. 作物病害监测预警研究进展[J]. 植物保护学报，49（1）：298-315.

金玉，2020. 槟榔黄化病多源遥感数据监测研究[D]. 合肥：安徽大学.

李艳玲，袁全国，任升，2023. 泡桐脱毒苗组培快繁技术研究[J]. 现代农业科技（22）：90-92，109.

刘仲健，罗焕亮，张景宁，1999. 植原体病理学[M]. 北京：中国林业出版社

罗大全，陈慕容，叶沙冰，等，2001. 海南槟榔黄化病的病原鉴定研究[J]. 热带作物学报，22（2）：43-46.

罗大全，车海彦，曹学仁，等，2016. 海南槟榔黄化病疫情监测网络信息平台的研究与应用：中国植物病理学会2016年学术年会论文集[C]. 北京：中国农业科学技术出版社.

孙圆龙，2022. 作物病害状态监测与预警技术研究[D]. 天津：河北工业大学.

田国忠，邓宝红，张兆欣，等，2009. 泡桐脱毒组培和规模化生产关键技术改进[J]. 林业科技开发，23（6）：73-78.

田国忠，李志清，胡佳续，等，2013. 我国部分枣树品种（系）的枣疯病抗性鉴定[J]. 林业科技开发，27（3）：19-25.

肖京，杨艳荣，赵锦，等，2013. 骏枣不同株系间的枣疯病抗性多样性[J]. 中国农业科学（23）：4977-4984.

苑晓伟，袁周伟，谭超，等，2020. 中国叶蝉寄生蜂的研究进展[J]. 贵州农业科学，48（1）：63-68.

张曙光，范怀忠，徐秀华，等，1999. 广东水稻橙叶病发病条件及防治研究[J]. 植物保护学报，26（3）：230-234.

赵锦，刘孟军，周俊义，等，2006. 抗枣疯病种质资源的筛选与应用[J]. 植物遗传资源学报，7（4）：398-403.

赵晋陵，金玉，叶回春，等，2020. 基于无人机多光谱影像的槟榔黄化病遥感监测[J]. 农业工程学报，36（8）：54-61.

ELDER R J，REID D J，MACLEOD W N B，et al.，2002. Post-ratoon growth and yield of three hybrid papayas（*Carica papaya L.*）under mulched and bare-ground conditions[J]. Australian Journal of Experimental Agriculture，42（1）：71-81.

WALSH K B，GUTHRIE J N，WHITE D T，2006. Control of phytoplasma diseases of papaya in Australia using netting[J]. Australasian Plant Pathology，35：49-54.

WEINTRAUB P G，JONES P，2010. Phytoplasmas：Genomes，Plant Hosts and Vectors[M]. Wallingford，UK：CABI.

第八章

展望

植原体是一类没有细胞壁、寄生于植物体内部细胞和昆虫体内的细菌，并能在二者之间通过媒介昆虫进行传播，引起植物病害。植原体是植物病理学研究领域中的一个重要方向，有着悠久的研究历史。近年来，关于植原体及其病害的研究主要侧重于检测诊断、寄主范围、植原体遗传特点及产生机制、植原体-寄主互作机制以及植原体病害绿色防控等方面研究。

一、检测诊断

由于分离培养的困难，人们对植原体形态、培养性状和生理代谢差异的了解甚少。因而血清学技术（Arashida et al., 2008；Wang et al., 2010）和分子生物学技术成为植原体遗传多样性研究的主要手段（Lee et al., 2004；Malembic-Maher et al., 2011）。植原体病害的检测通常包括症状观察、电子显微镜检测、组织化学染色法、血清学检测和分子生物学检测等方法。寄主植物在植原体感染后有时不显示典型症状，有时显示的症状难以与病毒或生理性病害区分。在电子显微镜下，植原体呈多态不规则形，具有双层膜结构，平均直径为200～800 nm（Bertaccini, 2007）。对植物组织切片的显微观察对于识别植原体很重要。然而，因为植原体的形态与寄主细胞器和杂质非常相似，有时观察植原体并不理想（Al-Zadjali et al., 2012；Khadhair et al., 2001）。作为一种致病微生物，植原体会导致韧皮部筛管沉积愈创木质素。使用苯胺蓝染色，在荧光显微镜下可以观察到病变组织中大量的荧光斑点，间接证明植原体的存在（De Marco et al., 2016；Musetti et al., 2010, 2013）。此外4′,6-二脒基-2苯基吲哚（DAPI）是一种特异性结合于富含A+T碱基的DNA的染料，如植原体基因组中的DNA。因此，DAPI染色可作为植原体基因组DNA的指示剂（Asudi et al., 2021；Dai et al., 1997；Thomas & Balasundaran, 1998）。然而，这些组织化学染色方法是间接的，而不是特异性染色植原体本身，容易出现假阳性结果。

使用血清学方法进行检测时，抗原-抗体组合具有极高的特异性。然而，由于韧皮部筛管中植原体数量少，纯化困难，这些方法的应用受到限制。最近报道了使用植原体效应蛋白的血清测定法，允许在具有明显症状的植物上识别特定植原体（Bai et al., 2009；2022）。此外，还有学者建立了一种制备探针的方法，这些探针可以用于核酸杂交分析，以便在寄主植物和传播媒介（即传播这些细菌的昆虫）中检测到这些植原体（Aldaghi et al., 2007；Nair et al., 2016；Ramjegathesh et al., 2019）。然而，对于实验室研究，推荐结合PCR和测序技术，并且可以基于测序进行特定分类。目前，基于植原体16S rDNA基因的实时荧光定量PCR（q-PCR）也已经开发（Hou et al., 2009）。与PCR相比，这种方法减少了染色和电泳步骤，只需1 h即可完成检测，并提高了检测效率（Hou et al., 2009）。该检测方法比巢式PCR更敏感（Jones et al., 2021），可以检测无症状或抗性组织中低浓度的植原体（Han et al., 2013）。

近年来，环介导等温扩增（LAMP）技术也是植原体检测中的热点之一（Ye et al., 2022）。然而，常见的LAMP检测特定于某些植原体类型或组，而且LAMP检测技术高度敏感，容易被核酸污染，导致程序中出现假阳性结果，仍需进一步优化相关工作（Ye et al., 2022；Zhou et al., 2022）。与PCR检测相比，基于CRISPR系统的检测技术可以避免核酸扩增（Gurr et al., 2016），这在最大程度上防止了样本间的交叉污染，检测结果更为准确。因此，建立植原体的CRISPR检测技术可能成为未来的研究方向。

二、寄主范围

近年来，植原体病害造成的经济危害变得越来越严重。据统计，全球范围内植原体病害的报道广泛存在，常见于温带和热带地区（Lee et al., 2000）。据估计，植原体已损害了超过1 000种植物（Hiruki & Wang, 2004；Seemüller & Schneider, 2004；Streten & Gibb, 2003），包括粮食作物、油料作物（Jaiswal et al., 2019；Thorat et al., 2016）、蔬菜、果树、园林植物、观赏植物、牧草和杂草等，造成了巨大的经济损失（Asudi et al., 2021；Bertaccini, 2007；Gurr et al., 2016；Maejima et al., 2014；Oshima et al., 2004）。小麦、水稻、玉米和大豆4种主要粮食作物都容易受到植原体感染。对植原体敏感的油料作物包括油菜、芝麻和花生；蔬菜包括番茄、辣椒、甘薯、马铃薯、胡萝卜和生菜；水果包括草莓、葡萄、苹果、梨、杏、李子、枣、橄榄和椰子；园林木材包括榆树、柳树、棕榈和泡桐；观赏植物包括仙人掌、长春花、菊花和丁香。植原体在不同寄主中常常引起不同的症状；例如，小麦蓝矮植原体在小麦上引起黄化、矮化和丛枝，但在长春花上引起黄化、矮化、丛枝、小叶和绿变（Wu et al., 2005）。植原体的中间寄主植物可能存在于作物附近，也可能是媒介昆虫获取植原体的来源（Wu et al., 2010）。确定植原体的寄主范围有助于通过打破植原体病害循环控制植原体病害，亦是未来植原体及其病害研究的热点之一。

三、植原体遗传特点及产生机制

在现今发现的众多微生物代谢体系中,植原体拥有最简单、规模最小的代谢体系和生命自我复制方式。植原体生活在营养丰富的植物韧皮部环境中,一些自身不能合成的物质可从寄主体内获取,因而对自身基因组的依赖性逐渐减少,在遗传上逐渐发生退化,在进化过程中丢掉了更多的基因,基因组逐渐变小,能够独立执行的代谢功能也逐渐减少,这种进化方式在遗传上被认为是独特的基因减少进化(reductive evolution)。但植原体保留了一些寄生生活必不可少的基因,如编码转运蛋白的基因。据推测,植原体中超过一半的蛋白质位于植原体原生质膜上,参与植原体原生质膜和寄主细胞之间的互作。另一方面,植原体寄居在寄主细胞内,和寄主细胞的原生质密切接触,在长期的进化过程中,面对着各种强大的、与其他细菌所不同的选择压力,植原体染色体之间、植原体同寄主染色体之间、植原体同介体昆虫体内的病毒之间都可能发生遗传物质重组或交换,这样一来,植原体基因组的进化就会产生丰富的遗传多样性(Nishigawa et al.,2002a,b;Liefting et al.,2004;Andersen et al.,2013)。

基于单一植原体基因组学信息的分析推断,显然无法系统阐明植原体确切的代谢网络,需要大量关键基因表达与调控、生化和代谢过程,乃至与表型相关性的实验和分析证据的支持(Kube et al.,2012)。进一步通过比较基因组学和功能基因组学研究揭示植原体有别于其他细菌及寄主植物的独特基因,特别是假基因的结构、组织、生理、生化和代谢调控过程细节,遗传进化的特点以及程度不同的重复序列形成机制和生物学意义,将不仅是基础研究的重点,而且有助于研发新的治疗药剂和制定更有效的病害防控对策(Tran-Nguyen et al.,2008)。例如,对洋葱黄化植原体的2个相邻操纵子S10和spc序列的分析发现,二者共用1个启动子,因而S10和spc基因簇可能作为1个S10-spc操纵子而起始转录,这种特性更类似于芽孢杆菌(Miyata et al.,2002)。苹果簇叶植原体的$fusA$和tuf基因位于$rps7$的下游且紧密连接在一起,而可培养的生殖道支原体(Mycoplasma genitalium)和肺炎支原体(M. pneumonia)$fusA$和tuf基因则分别位于不同的位点而分别转录(Berg & Seemüller,1999)。

随着基因组测序技术的发展,全基因组核苷酸序列一致性已成为评估物种界限和估计2个基因组之间遗传关系的有力支撑。通过各种测序技术和比较基因组组装,到目前为止已发布了27个完整基因组和216个草图基因组(https://www.ncbi.nlm.nih.gov/datasets)。这些测序结果表明,植原体的G+C含量不超过30%,且最小值为19.9%。对植原体基因组编码基因的分析揭示了植原体中的基本代谢途径,包括参与DNA复制、转录、翻译和蛋白质转移的途径(Kube et al.,2014)。植原体缺乏许多重要的代谢能力,包括蛋白质和脂类生物合成、三羧酸循环、氧化磷酸化、戊糖磷酸途径和ATP合酶(Oshima et al.,2004)。植原体虽然代谢能力较弱,但其基因组中含有许多编码转运蛋白的基因,这些转运蛋白能够从寄主细胞吸收糖类、氨基酸、寡肽和无机盐，表明植原体高度依赖寄主代谢物（Kube et al., 2014），然而分泌蛋白的数量并不随基因组大小增加，因此，植原体全基因组研究有待进一步深入探索。

四、植原体–寄主互作机制研究

植原体的代谢能力较弱，但它们的转运和分泌系统相对活跃（Chen et al., 2014；Saccardo et al., 2012）。大多数植原体预计含有十几个或几十个分泌蛋白，这些蛋白很可能作为效应蛋白发挥作用（Oshima et al., 2004；Wei et al., 2022）。近年来，许多效应蛋白的功能已被揭示。这些植原体效应蛋白的功能可以分为2类。一类引起丛枝、叶片卷曲、花绿变、矮化、不育、调节植物激素和挥发物水平、调控植物防御反应和破坏细胞完整性，这些效应蛋白的功能是调节植物的生长和发育；另一类则通过促进昆虫的吸引和繁殖功能，来调节植物与昆虫之间的相互作用。同一植原体的不同效应蛋白可以协同工作，造成类似的植物症状。例如，AYWB效应蛋白SAP11通过介导TCP转录因子的不稳定性，刺激腋芽的增殖（Sugio et al., 2014；Zhou et al., 2021）。效应蛋白SAP05通过泛素非依赖途径介导SPL和GATA发育调节因子的同时降解，诱导叶片和不育分枝的增加，并延长寄主寿命（Huang et al., 2021）。SAP54可以通过与RAD23家族蛋白相互作用，介导MADS-box转录因子（MTFs）的降解，将花朵转变为叶状植物组织并诱导植物不育（Iwabuchi et al., 2019）。同一效应蛋白可能在不同的植原体中可能发挥相同或不同的作用，这可能是由于不同寄主和植原体的持续进化过程。例如，SAP11、SWP1、SJP1和SJP2需要N末端核定位信号和C末端卷曲螺旋结构域才能使TCP不稳定，导致丛枝病（Al-Subhi et al., 2021；Chang et al., 2018；Sugio, Kingdom et al., 2011；Wang et al., 2018；Zhou et al., 2021）。此外，PaWB的SAP54在寄主植物中诱发丛枝病，而AY-WB的SAP54在寄主中引起花绿变（Cao et al., 2021；MacLean et al., 2014）。诱发花变叶的植原体效应蛋白家族与花发育同源MTFs相互作用，降解SEP3和AP1，导致下游花发育基因的异常表达和叶状花的形成（Iwabuchi et al., 2020；Maejima et al., 2014）。不同植原体的不同效应蛋白可能具有相似的功能。例如，SAP11AY-WB和SAP54PaWB可以诱导丛枝病，而SAP11AY-WB作用于TCP，但SAP54PaWB作用于SPLa（Cao et al., 2021；Sugio et al., 2011）。植原体效应蛋白同源物的功能存在变化，主要是由于关键位点的氨基酸变异。由称为phyllogens的植原体效应蛋白家族引起的花器官畸形，特别是花绿变，是植原体感染的常见症状（Iwabuchi et al., 2020）。Phyllogens是植原体中保守的蛋白质家族。Phyllogens通过与寄主蛋白相互作用并诱导其降解，这些蛋白是花发育所必需的，如MADS结构域转录因子（Kitazawa et al., 2022）。系统发育分析将phyllogens分为4组，包括phyl-A、phyl-B、phyl-C和phyl-D。phyl-B组phyllogens中的单个氨基酸多态性阻止了

phyllogens与A-和E-类MTFs的相互作用（Iwabuchi et al.，2020）。结果表明，花器官不再转变为叶状结构（Iwabuchi et al.，2020）。诱导剂（TENGU）诱发丛枝病和矮化病，是一种小型分泌蛋白（Hoshi et al.，2009）。SWP12是由小麦蓝矮植原体分泌的潜在效应蛋白，可调控寄主的生理反应以创造有利于植原体定殖的环境（Bai et al.，2022）。由此可见，效应蛋白功能的研究显著地有助于理解植原体病害症状形成的机制，是未来控制植原体病害的有效途径。然而，植原体感染引起的黄化、矮化和小叶形成的机制仍然理解不足，有待进一步探索。

植原体作为韧皮部寄生体，需要从植物或昆虫寄主中获取养分，并引起植物生理和病理上的变化。目前，对于植原体感染所引起的黄化和衰败症状的理解不足，而植原体导致矮化症状的具体机制也未确定。效应蛋白的靶点是关键的植物蛋白，其中大多数是转录因子，所有这些蛋白都通过26S蛋白酶体途径被降解。与植原体相关的效应蛋白直接影响泛素-蛋白酶体系统的组成部分或作为该系统的组成部分。利用植物26S蛋白酶体途径获得的植原体效应蛋白的功能及其对植物生理反应的后续影响仍需进一步研究。进一步探索植物生理和病理之间的平衡可能有助于阐明植原体病害的机制，理解效应蛋白的功能和作用机制对未来预防和控制这些病害至关重要。

五、绿色防控

尽管植原体的培养有一定难度，但人们已经确定了传播植原体的主要介体（叶蝉、木虱等）。通过植物的无性繁殖和嫁接也可以成功地在寄主植物上维持植原体。分子生物技术的进步极大地促进了对植原体的研究。候选基因组测序显示，植原体缺乏基础代谢可能是植物寄生性病害的原因。目前，已研究了多个效应蛋白的功能，进而揭示植原体病害的症状形成和定殖机制。这些结果不仅能洞察细胞内寄生体的生存模式，还将为植原体病害的预防和控制提供有价值的见解。由于植原体病害往往是致命的或具有破坏性损伤，因此植原体病害的早期识别和诊断可以促进控制措施的快速发展和实施。通过调查某一地区的植原体物种，可以利用基因组信息构建广谱和特定病害的检测系统。基于每个地理区域现有植原体的分布，如果该区域没有植原体，可以将某个植原体物种归类为入境检疫对象。根据植原体的类型，可以分析主要寄主植物，并为该地区重要经济植物制定预防措施。应使用并优化现有的预防和控制方法，以防止病害的传播并最小化病害造成的直接经济损失。除了防止传播外，治疗受损植物是未来植原体研究的一个挑战。基于目前对病理形成机制的理解，可以探索病情恢复的绿色防控策略。

现有的四环素类抗生素药剂治疗枣疯病过程中存在较严重的植物毒性，而治疗效果不彻底及加入WTO后可能出现新的食品安全等问题，又进一步限制了该类药剂的应用范围。选育和栽种抗病品种是控制这一病害最为经济有效和根本的措施。近年来，我国在抗植原体植物品种选育方面已有了很大的进展，如在枣树品种的抗病性测定方面，有近百份枣树种质资源得到了抗病能力鉴定，并从中筛选出了高抗枣疯病的品系。同时也对一些高抗品种的抗病机制进行了研究，包括对水杨酸、茉莉酸、蛋白质及DNA等分子标记等方面的探讨（温秀军等，2005；刘孟军等，2009；田国忠等，2013；于少帅，2016）。在植物病害防治过程中，化学农药的大量使用造成了植物农药残留、人畜中毒、环境污染等一系列严重问题。因此，植物天然活性产物的开发和利用越来越受到人们的重视（赵良忠等，2004；曾超珍等，2009）。植物或微生物源抗植原体相关活性物质的筛选及其作用机理研究不仅对于阐明植物抗植原体机制具有重要意义，也会对治疗植原体病害绿色药剂的开发产生重要的促进作用。这些研究的深入开展将会极大地提高植原体病害防治技术水平。

参考文献

刘孟军，赵锦，周俊义，2009. 枣疯病[M]. 北京：中国农业出版社.

田国忠，李志清，胡佳续，等，2013. 我国部分枣树品种（系）的枣疯病抗性鉴定[J]. 林业科技开发，27（3）：19-25.

温秀军，郭晓军，田国忠，等，2005. 几个枣树品种和婆枣单株对枣疯病抗性的鉴定[J]. 林业科学，41（3）：88-96.

于少帅，2016. 植原体 *tuf* 基因启动子分子特征和枣树抗植原体物质研究[D]. 北京：中国林业科学研究院.

曾超珍，刘志祥，2009. 鹅掌楸抑菌物质的体外抗菌活性及稳定性试验[J]. 湖北农业科学，48（5）：1222-1224.

赵良忠，王放银，段林东，2004. 大青叶抗菌物质提取及抗菌效果研究[J]. 食品科学，25（11）：138-140.

ALDAGHI M，MASSART S，ROUSSEL S，et al.，2007. Development of a new probe for specific and sensitive detection of 'Candidatus phytoplasma mali' in inoculated apple trees[J]. Annals of Applied Biology，151：251-258.

AL-SUBHI A M，AL-SADI A M，AL-YAHYAI R A，et al.，2021. Witches'-broom disease of lime contributes to phytoplasma epidemics and attracts insect vectors[J]. Plant Disease，105：2637-2648.

AL-ZADJALI A D，AL-SADI A M，DEADMAN M L，et al.，2012. Detection，identification and molecular characterization of a phytoplasma associated with beach naupaka witches'-broom[J]. Journal of Plant Pathology，94：379-385.

ANDERSEN M T，LIEFTING L W，HAVUKKALA I，et al.，2013. Comparison of the complete genome sequence of two closely related isolates of 'Candidatus phytoplasma australiense' reveals genome plasticity[J]. BMC Genomics, 14: 529.

ARASHIDA R, KAKIZAWA S, ISHII Y, et al., 2008. Cloning and characterization of the antigenic membrane protein (Amp) gene and in situ detection of Amp from malformed flowers infected with Japanese hydrangea phyllody phytoplasma[J]. Phytopathology, 98: 769-775.

ASUDI G O, OMENGE K M, PAULMANN M K, et al., 2021. The physiological and biochemical effects on Napier grass plants following napier grass stunt phytoplasma infection[J]. Phytopathology, 111: 703-712.

BAI B, ZHANG G, LI Y, et al., 2022. The 'Candidatus Phytoplasma tritici' effector SWP12 degrades the transcription factor TaWRKY74 to suppress wheat resistance[J]. The Plant Journal, 112: 1473-1488.

BAI X, CORREA V R, TORUNO T Y, et al., 2009. AY-WB phytoplasma secretes a protein that targets plant cell nuclei[J]. Molecular Plant-Microbe Interactions, 22: 18-30.

BERG M, SEEMÜLLER E, 1999. Chromosomal organization and nucleotide sequence of the elongation factors G and Tu of the apple proliferation phytoplasma[J]. Gene, 226: 103-109.

BERTACCINI A, 2007. Phytoplasmas: diversity, taxonomy, and epidemiology[J]. Frontiers in Bioscience-Landmark, 12: 673-689.

CAO Y, SUN G, ZHAI X, et al., 2021. Genomic insights into the fast growth of paulownias and the formation of Paulownia witches'-broom[J]. Molecular Plant, 14: 1668-1682.

CHANG S H, TAN C M, WU C T, et al., 2018. Alterations of plant architecture and phase transition by the phytoplasma virulence factor SAP11[J]. Journal of Experimental Botany, 69: 5389-5401.

CHEN W, LI Y, WANG Q, et al., 2014. Comparative genome analysis of wheat blue dwarf phytoplasma, an obligate pathogen that causes wheat blue dwarf disease in China[J]. PLoS One, 9: e96436.

DAI Q, HEM F T, LIU P Y, 1997. Elimination of phytoplasma by stem culture from mulberry plants (Morus alba) with dwarf disease[J]. Plant Pathology, 46: 56-61.

DE MARCO F, PAGLIARI L, DEGOLA F, et al., 2016. Combined microscopy and molecular analyses show phloem occlusions and cell wall modifications in tomato leaves in response to 'Candidatus Phytoplasma solani'[J]. Journal of Applied Microbiology, 263: 212-225.

GURR G M, JOHNSON A C, ASH G J, et al., 2016. Coconut lethal yellowing diseases: a phytoplasma threat to palms of global economic and social significance[J]. Frontiers in Plant

Science, 7: 1521.

HAN S S, LEE K J, KAMALA-KANNAN S, 2013. Detection of aster yellows phytoplasma (16SrⅠ) associated with prickly-ash (*Zanthoxylum schinifolium* S. et Z.) witches broom disease in Korea[J]. Journal of Phytopathology, 161: 582-585.

HIRUKI C, WANG K, 2004. Clover proliferation phytoplasma: '*Candidatus* Phytoplasma trifolii'[J]. International Journal of Systematic and Evolutionary Microbiology, 54: 1349-1353.

HOU W, WU K, LI Y, et al., 2009. Isolation and bioinformatics analysis of hemolysin gene from wheat blue dwarf phytoplasma[J]. Journal of Agricultural Biotechnology, 17: 665-669.

HOSHI A, OSHIMA K, KAKIZAWA S, et al., 2009. A unique virulence factor for proliferation and dwarfism in plants identified from a phytopathogenic bacterium[J]. Proceedings of the National Academy of Sciences of the United States of America, 106: 6416-6421.

HUANG W, MACLEAN A, SUGIO A, et al., 2021. Parasitic modulation of host development by ubiquitin independent protein degradation[J]. Cell, 184: 5201-5214.

IWABUCHI N, MAEJIMA K, KITAZAWA Y, et al., 2019. Crystal structure of phyllogen, a phyllodyinducing effector protein of phytoplasma[J]. Biochemical and Biophysical Research Communications, 513: 952-957.

JAISWAL S, JADHAV P V, JASROTIA R S, et al., 2019. Transcriptomic signature reveals mechanism of flower bud distortion in witches'-broom disease of soybean (*Glycine max*)[J]. BMC Plant Biology, 19: 26-37.

JONES L M, PEASE B, PERKINS S L, et al., 2021. '*Candidatus* Phytoplasma dypsidis', a novel taxon associated with a lethal wilt disease of palms in Australia[J]. International Journal of Systematic and Evolutionary Microbiology, 71: 004818.

KHADHAIR A H, TEWARI J P, HOWARD R J, et al., 2001. Detection of aster yellows phytoplasma in false flax based on PCR and RFLP[J]. Microbiological Research, 156: 179-184.

KITAZAWA Y, IWABUCHI N, MAEJIMA K, et al., 2022. A phytoplasma effector acts as a ubiquitinlike mediator between floral MADS-box proteins and proteasome shuttle proteins[J]. The Plant Cell, 34: 1709-1723.

KUBE M, MITROVIC J, DUDUK B, et al., 2012. Current view on phytoplasma genomes and encoded metabolism[J]. The Scientific World Journal, article ID 185942.

KUBE M, SIEWERT C, MIGDOLL A M, et al., 2014. Analysis of the complete genomes of *Acholeplasma brassicae*, *A. palmae* and *A. laidlawii* and their comparison to the obligate parasites from 'Candidatus Phytoplasma' [J]. Journal of Molecular Microbiology and Biotechnology, 24: 19-36.

LEE I M, DAVIS R E, GUNDERSEN-RINDAL D E, 2000. Phytoplasma: phytopathogenic mollicutes[J]. Annual Review of Microbiology, 54: 221-255.

LEE I M, GUNDERSEN-RINDAL D E, DAVIS R E, et al., 2004. 'Candidatus Phytoplasma asteris', a novel phytoplasma taxon associated with aster yellows and related diseases[J]. International Journal of Systematic and Evolution Microbiology, 54: 1037-1048.

LIEFTING L W, SHAW M E, KIRKPATRIC B C, 2004. Sequence analysis of two plasmids from the phytoplasma beet leafhopper-transmitted virescence agent[J]. Microbiology, 150: 1809-1817.

MACLEAN A M, ORLOVSKIS Z, KOWITWANICH K, et al., 2014. Phytoplasma effector SAP54 hijacks plant reproduction by degrading MADS-box proteins and promotes insect colonization in a RAD23-dependent manner[J]. PLoS Biology, 12: e1001835.

MAEJIMA K, IWAI R, HIMENO M, et al., 2014. Recognition of floral homeotic MADS domain transcription factors by a phytoplasmal effector, phyllogen, induces phyllody[J]. The Plant Journal, 78: 541-554.

MALEMBIC-MAHER S, SALAR P, FILIPPIN L, et al., 2011. Genetic diversity of European phytoplasmas of the 16SrⅤ taxonomic group and proposal of 'Candidatus Phytoplasma rubi'[J]. International Journal of Systematic and Evolutionary Microbiology, 61: 2129-2134.

MIYATA S, FURUKI K, OSHIMA K, et al., 2002. Complete nucleotide sequence of the S10-spc operon of phytoplasma: gene organization and genetic code resemble those of Bacillus subtilis[J]. DNA and Cell Biology, 21: 527-534.

MUSETTI R, BUXA S V, DE MARCO F, et al., 2013. Phytoplasma-triggered Ca^{2+} influx is involved in sieve tube blockage[J]. Molecular Plant-Microbe Interactions, 26: 379-386.

MUSETTI R, PAOLACCI A, CIAFFI M, et al., 2010. Phloem cytochemical modification and gene expression following the recovery of apple plants from apple proliferation disease[J]. Phytopathology, 100: 390-399.

NAIR S, MANIMEKALAI R, SOUMYA V P, et al., 2016. Dual labeled probe based real time PCR method for detection of 16SrⅪ-B sub-group phytoplasma associated with coconut root wilt disease in India[J]. Australasian Plant Pathology, 45: 187-189.

NISHIGAWA H, OSHIMA K, KAKIZAWA S, et al., 2002a. A plasmid from a non-insect-transmissible line of a phytoplasma lacks two open reading frames that exist in the plasmid from the wild-type line[J]. Gene, 298: 195-201.

NISHIGAWA H, OSHIMA K, KAKIZAWA S, et al., 2002b. Evidence of intermolecular recombination between extrachromosomal DNAs in phytoplasma: a trigger for the biological diversity of phytoplasma[J]. Microbiology, 148: 1389-1396.

OSHIMA K, KAKIZAWA S, NISHIGAWA H, et al., 2004. Reductive evolution suggested from the complete genome sequence of a plant-pathogenic phytoplasma[J]. Nature Genetics, 36: 27-29.

RAMJEGATHESH R, KARTHIKEYAN G, BALACHANDAR D, et al., 2019. Nested and TaqMan® probe based quantitative PCR for the diagnosis of Ca. Phytoplasma in coconut palms[J]. Molecular Biology Reports, 46: 479-488.

SACCARDO F, MARTINI M, PALMANO S, et al., 2012. Genome drafts of four phytoplasma strains of the ribosomal group 16SrⅢ[J]. Microbiology (Reading England), 158: 2805-2814.

SEEMÜLLER E, SCHNEIDER B, 2004. 'Candidatus Phytoplasma mali', 'Candidatus Phytoplasma pyri' and 'Candidatus Phytoplasma prunorum', the causal agents of apple proliferation, pear decline and European stone fruit yellows, respectively[J]. International Journal of Systematic and Evolutionary Microbiology, 54: 1217-1226.

SUGIO A, KINGDOM H N, MACLEAN A M, et al., 2011. Phytoplasma protein effector SAP11 enhances insect vector reproduction by manipulating plant development and defense hormone biosynthesis[J]. Proceedings of the National Academy of Sciences of the United States of America, 108: E1254-E1263.

STRETEN C, GIBB K S, 2003. Identification of genes in the tomato big bud phytoplasma and comparison to those in sweet potato little leaf-V4 phytoplasma[J]. Microbiology, 149: 1797-1805.

THOMAS S, BALASUNDARAN M, 1998. In situ detection of phytoplasma in spike-disease-affected sandal using DAPI stain[J]. Current Science, 74: 989-993.

THORAT V, MORE V, JADHAV P, et al., 2016. First report of a 16SrⅡ-D group phytoplasma associated with witches'-broom disease of soybean (Glycine max) in Maharashtra, India[J]. Plant Disease, 100: 2521-2522.

TRAN-NGUYEN L T T, KUBE M, SCHNEIDER B, et al., 2008. Comparative genome analysis of 'Candidatus phytoplasma australiense' (subgroup tuf-Australia I; rp-A) and 'Ca. phytoplasma asteris' strains OY-M and AY-WB[J]. Journal of Bacteriology, 190: 3979-3991.

WANG J, ZHU X P, GAO R, et al., 2010. Genetic and serological analyses of elongation factor EF-Tu of paulownia witches'-broom phytoplasma（16Sr I -D）[J]. Plant Pathology, 59：972-981.

WANG N, YANG H, YIN Z, et al., 2018. Phytoplasma effector SWP1 induces witches'-broom symptom by destabilizing the TCP transcription factor BRANCHED1[J]. Molecular Plant Pathology, 19：2623-2634.

WEI W, INABA J, ZHAO Y, et al., 2022. Phytoplasma infection blocks starch breakdown and triggers chloroplast degradation, leading to premature leaf senescence, sucrose reallocation, and spatiotemporal redistribution of phytohormones[J]. International Journal of Molecular Sciences, 23：1810.

WU Y, GU P, AN F, et al., 2005. Host range of wheat blue dwarf phytoplasma[J]. Journal of Northwest Sci-Tech University of Agriculture and Forestry（Natural Science Edition）, 33：8-10.

WU Y, HAO X, LI Z, et al., 2010. Identification of the phytoplasma associated with wheat blue dwarf disease in China[J]. Plant Disease, 94：977-985.

YE H, NOWAK C, LIU Y, et al., 2022. Plasmonic LAMP：improving the detection specificity and sensitivity for SARS-CoV-2 by plasmonic sensing of isothermally amplified nucleic acids[J]. Small, 18：e2107832.

ZHOU J, MA F, YAO Y, et al., 2021. Jujube witches'-broom phytoplasma effectors SJP1 and SJP2 induce lateral bud outgrowth by repressing the ZjBRC1-controlled auxin efflux channel[J]. Plant, Cell and Environment, 44：3257-3272.

ZHOU X, SCHUH D A, CASTLE L M, et al., 2022. Recent advances in signal amplification to improve electrochemical biosensing for infectious diseases[J]. Frontiers in Chemistry, 10：911678.